ESSENTIALS OF NUMBER THEORY

Edited by
Maria Catherine C. Borres

Arcler Press

Essentials of Number Theory

Edited by Maria Catherine C. Borres

Edition 2018

ArclerPress

Published by
Arcler Press LLC
708 3rd Avenue, 6th Floor
New York
NY 10017
United States of America
www.arclerpress.com

ISBN: 978-1-68094-586-7 (Hardback)

Printed in United States of America

Library of Congress Control Number: 2017933173

About the Editor

Maria Catherine C. Borres

Catherine is currently taking up Master of Arts in Education Major in Mathematics in Philippine Normal University – Manila. She is currently working as a Content Developer for Mathematics at the Affordable Private Education Center (APEC Schools).

Table of Contents

2 NUMBER THEORY: AN OVERVIEW

3 ALGEBRAIC NUMBER THEORY

4 ANALYTIC NUMBER THEORY

5 GEOMETRY OF NUMBERS

6 COMPUTATIONAL NUMBER THEORY

7 TRANSCENDENTAL NUMBER THEORY

Preface

Number theory is a vast and fascinating field of mathematics, sometimes called "higher arithmetic," consisting of the study of the properties of whole numbers. Primes and prime factorization are especially important in number theory, as are a number of functions such as the divisor function, Riemann zeta function, and totient function. Number theory has always fascinated amateurs as well as professional mathematicians. In contrast to other branches of mathematics, many of the problems and theorems of number theory can be understood by laypersons, although solutions to the problems and proofs of the theorems often require a sophisticated mathematical background. Until the mid-20th century, number theory was considered the purest branch of mathematics, with no direct applications to the real world. The advent of digital computers and digital communications revealed that number theory could provide unexpected answers to real-world problems. At the same time, improvements in computer technology enabled number theorists to make remarkable advances in factoring large numbers, determining primes, testing conjectures, and solving numerical problems once considered out of reach. Modern number theory is a broad subject that is classified into subheadings such as elementary number theory, algebraic number theory, analytic number theory, geometric number theory, and probabilistic number theory. These categories reflect the methods used to address problems concerning the integers. Number theorists study prime numbers as well as the properties of objects made out of integers (e.g., rational numbers) or defined as generalizations of the integers (e.g., algebraic integers). Integers can be considered either in themselves or as solutions to equations (Diophantine geometry). Questions in number theory are often best understood through the study of analytical objects (e.g., the Riemann zeta function) that encode properties of the integers, primes or other number-theoretic objects in some fashion (analytic number theory). One may also study real numbers in relation to rational numbers, e.g., as approximated by the latter (Diophantine approximation).

Content Coverage

Chapters One and *Two* present an overview of numbers and number theory, correspondingly. A number is a theoretical concept, and a numeral is the way that people denote that concept. A numeral contains one or more written symbols, but a number can be expressed in a range of ways. Number and numeral are often used interchangeably, but there is a slight difference between them.

Chapters Three and *Four* present a coverage of algebraic numbers and analytic number theory, consistently. Algebraic number theory uses the techniques of abstract algebra to study the integers, rational numbers, and their generalizations, however analytic number theory uses methods from mathematical analysis to solve problems about the integers.

Chapters Five and *Six* are intended to focus on geometry of numbers and computational number theory, separately. In number theory, the geometry of numbers studies convex bodies and integer vectors in n-dimensional space, while computational number theory is concerned with finding and implementing efficient computer algorithms for solving various problems in number theory.

Chapter Seven explains the fundamentals of transcendental number theory. Transcendental numbers are important in the history of mathematics because their investigation provided the first proof that circle squaring, one of the geometric problems of antiquity that had baffled mathematicians for more than 2000 years was, in fact, insoluble.

CHAPTER 1

Introduction to Number

Chapter Contents

- **Introduction**
- **Numerals**
- **Main Classification**
- **History**
- **Natural Number**
- **Prime Number**
- **Rational Number**
- **Irrational Number**
- **Algebraic Number**
- **Transcendental Number**

A number is a theoretical concept, and a numeral is the way that people denote that concept. A numeral contains one or more written symbols, but a number can be expressed in a range of ways. Number and numeral are often used interchangeably, but there is a slight difference between them.

For instance, the number 5 is a concept used to describe five objects. One could express this concept by holding up five fingers or by saying the word "five." However, if one wrote down the number "5" that would be a numeral. Regardless of whether the person wrote that number using Roman (V) or Arabic (5) numerals or even letters (five), it would still be considered a numeral once it was written.

Any written number can be a numeral, regardless of how many symbols or digits it takes to create that number. For instance, "5" is a numeral but so is "505" and "5,678,987." However, the word numeral cannot be used interchangeably with digit. A digit is a single symbol used to make a number or numeral. The number 5 can be written with just one symbol or digit, while the number 505 needs three digits and 5,678,987 needs seven digits.

INTRODUCTION

A number is a mathematical object used to count, measure, and label. The original examples are the natural numbers 1, 2, 3, and so forth. A notational symbol that represents a number is called a numeral. In addition to their use in counting and measuring, numerals are often used for labels (as with telephone numbers), for ordering (as with serial numbers), and for codes (as with ISBNs). In common usage, number may refer to a symbol, a word, or a mathematical abstraction.

In mathematics, the notion of number has been extended over the centuries to include 0, negative numbers, rational numbers such as 1/2 and -2/3, real numbers such as $\sqrt{2}$ and π, and complex numbers, which extend the real numbers by adding a square root of -1. Calculations with numbers are done with arithmetical operations, the most familiar being addition, subtraction, multiplication, division, and exponentiation. Their study or usage is called arithmetic. The same term may also refer to number theory, the study of the properties of numbers.

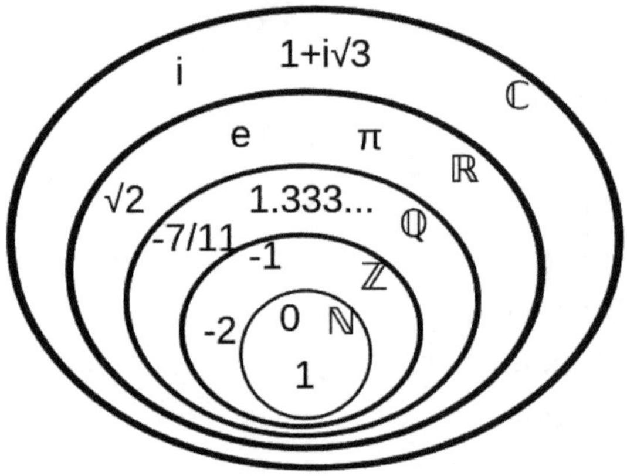

Subsets of the complex numbers.

Besides their practical uses, numbers have cultural significance throughout the world. For example, in Western society the number 13 is regarded as unlucky, and "a million" may signify "a lot." Though it is now regarded as pseudoscience, numerology, the belief in a mystical significance of numbers, permeated ancient and medieval thought. Numerology heavily influenced the development of Greek mathematics, stimulating the investigation of many problems in number theory which are still of interest today.

During the 19th century, mathematicians began to develop many different abstractions which share certain properties of numbers and may be seen as extending the concept. Among the first were the hypercomplex numbers, which consist of various extensions or modifications of the complex number system. Today, number systems are considered important special examples of much more general categories such as rings and fields, and the application of the term "number" is a matter of convention, without fundamental significance.

NUMERALS

Numbers should be distinguished from numerals, the symbols used to represent numbers. Boyer showed that Egyptians created the first ciphered numeral system. Greeks followed by mapping their counting numbers onto Ionian and Doric alphabets. The number five can be represented by digit "5" or by the Roman numeral "V". Notations used to represent numbers are discussed in the article numeral systems. An important development in the history of numerals was the development of a positional system, like modern decimals, which have many advantages, such as representing large numbers with only a few symbols. The Roman numerals require extra symbols for larger numbers.

MAIN CLASSIFICATION

Different types of numbers have many different uses. Numbers can be classified into sets, called **number systems**, such as the natural numbers and the real numbers. The same number can be written in many different ways. For different methods of expressing numbers with symbols, such as the Roman numerals.

Main number systems

$\mathbb{N}$	**Natural**	0, 1, 2, 3, 4, ... **or** 1, 2, 3, 4, ... $\mathbb{N}_0$ or $\mathbb{N}_1$ are sometimes used.
$\mathbb{Z}$	**Integer**	..., −5, −4, −3, −2, −1, 0, 1, 2, 3, 4, 5, ...
$\mathbb{Q}$	**Rational**	$\frac{a}{b}$ where a and b are integers and b is not 0
$\mathbb{R}$	**Real**	The limit of a convergent sequence of rational numbers
$\mathbb{C}$	**Complex**	$a + bi$ where a and b are real numbers and i is a formal square root of -1

Each of these number systems may be considered as a proper subset of the next one. This means that each one is canonically isomorphic to a proper subset of the next one, and that there is generally no problem with the abuse of notation consisting of identifying each number system with a subset of the next one. This is expressed, symbolically, by writing

$$\mathbb{N} \subset \mathbb{Z} \subset \mathbb{Q} \subset \mathbb{R} \subset \mathbb{C}.$$

Natural Numbers

The most familiar numbers are the natural numbers (sometimes called whole numbers or counting numbers): 1, 2, 3, and so on. Traditionally, the sequence of natural numbers started with 1 (0 was not even considered a number for the Ancient Greeks.) However, in the 19th century, set theorists and other mathematicians started including 0 (cardinality of the empty set, i.e. 0 elements, where 0 is thus the smallest cardinal number) in the set of natural numbers. Today, different mathematicians use the term to describe both sets, including 0 or not. The mathematical symbol for the set of all natural numbers is **N**, also written $\mathbb{N}$, and sometimes $\mathbb{N}_0$ or $\mathbb{N}_1$when it is necessary to indicate whether the set should start with 0 or 1, respectively.

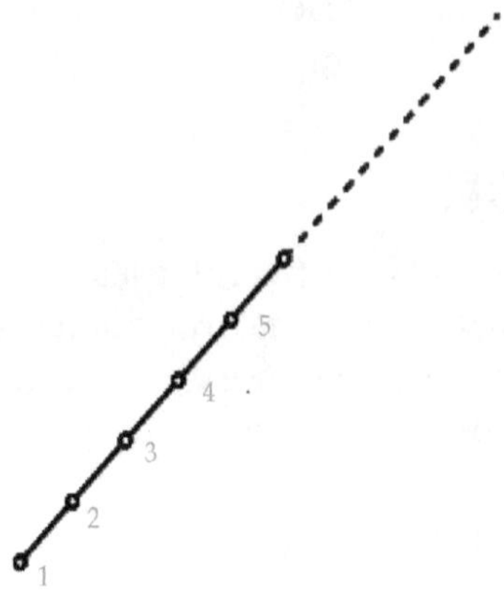

The natural numbers, starting with 1

In the base 10 numeral system, in almost universal use today for mathematical operations, the symbols for natural numbers are written using ten digits: 0, 1, 2, 3, 4, 5, 6, 7, 8, and 9. The radix or base is the number of unique numerical digits, including zero, that a numeral system uses to represent numbers (for the decimal system, the radix is 10). In this base 10 system, the rightmost digit of a natural number has a place value of 1, and every other digit has a place value ten times that of the place value of the digit to its right.

In set theory, which is capable of acting as an axiomatic foundation for modern mathematics, natural numbers can be represented by classes of equivalent sets. For instance, the number 3 can be represented as the class of all sets that have exactly three elements. Alternatively, in Peano Arithmetic, the number 3 is represented as sss0, where s is the "successor" function (i.e., 3 is the third successor of 0). Many different representations are possible; all that is needed to formally represent 3 is to inscribe a certain symbol or pattern of symbols three times.

Integers

The negative of a positive integer is defined as a number that produces 0 when it is added to the corresponding positive integer. Negative numbers are usually written with a negative sign (a minus sign). As an example, the negative of 7 is written −7, and 7 + (−7) = 0. When the set of negative numbers is combined with the set of natural numbers (including 0), the result is defined as the set of integers, **Z** also written $\mathbb{Z}$. Here the letter Z comes from German Zahl, meaning "number". The set of integers forms a ring with the operations addition and multiplication.

The natural numbers form a subset of the integers. As there is no common standard for the inclusion or not of zero in the natural numbers, the natural numbers without zero are commonly referred to as positive integers, and the natural numbers with zero are referred to as non-negative integers.

Rational Numbers

A rational number is a number that can be expressed as a fraction with an integer numerator and a positive integer denominator. Negative denominators are allowed, but are commonly avoided, as every rational number is equal to a fraction with positive denominator. Fractions are written as two integers, the numerator and the denominator, with a dividing bar between them. The fraction m/n represents m parts of a whole divided into n equal parts. Two different fractions may correspond to the same rational number; for example 1/2 and 2/4 are equal, that is:

$$\frac{1}{2} = \frac{2}{4}.$$

If the absolute value of m is greater than n (supposed to be positive), then the absolute value of the fraction is greater than 1. Fractions can be greater than, less than, or equal to 1 and can also be positive, negative, or 0. The set of all rational numbers includes the integers, since every integer can be written as a fraction with denominator 1. For example −7 can be written −7/1. The symbol for the rational numbers is **Q** (for quotient), also written $\mathbb{Q}$.

Real Numbers

The real numbers include all the measuring numbers. The symbol for the real numbers is **R**, also written as $\mathbb{R}$. Real numbers are usually represented by using decimal numerals, in which a decimal point is placed to the right of the digit with place value 1. Each digit to the right of the decimal point has a place value one-tenth of the place value of the digit to its left. For example, 123.456 represents 123456/1000, or, in words, one hundred, two tens, three ones, four tenths, five hundredths, and six thousandths. A finite decimal representation allows us to represent exactly only the integers and those rational numbers whose denominators have only prime factors which are factors of ten. Thus one half is 0.5, one fifth is 0.2, one tenth is 0.1, and one fiftieth is 0.02. To represent the rest of the real numbers requires an infinite sequence of digits after the decimal point. Since it is impossible to write infinitely many digits, real numbers are commonly represented by rounding or truncating this sequence, or by establishing a pattern, such as 0.333..., with an ellipsis to indicate that the pattern continues. Thus 123.456 is an approximation of any real number between 1234555/10000 and 1234565/10000 (rounding) or any real number between 123456/1000 and 123457/1000 (truncation). Negative real numbers are written with a preceding minus sign: -123.456.

Every rational number is also a real number. It is not the case, however, that every real number is rational. A real number, which is not rational, is called irrational. A decimal represents a rational number if and only if it has a finite number of digits or eventually repeats for ever, after any initial finite string digits. For example, 1/2 = 0.5 and 1/3 = 0.333... (forever repeating 3s, otherwise written 0.3). On the other hand, the real number π, the ratio of the circumference of any circle to its diameter, is

$$\pi = 3.14159265358979\ldots$$

Since the decimal neither ends nor eventually repeats forever (see: proof that pi is irrational) it cannot be written as a fraction, and is an example of an irrational number. Other irrational numbers include

$$\sqrt{2} = 1.41421356237\ldots$$

(the square root of 2, that is, the positive number whose square is 2).

Just as the same fraction can be written in more than one way, the same decimal may have more than one representation. 1.0 and 0.999... are two different decimal numerals representing the natural number 1. There are infinitely many other ways of representing the number 1, for example 1.00, 1.000, and so on.

Every real number is either rational or irrational. Every real number corresponds to a point on the number line. The real numbers also have an important but highly technical property called the least upper bound property.

When a real number represents a measurement, there is always a margin of error. This gives rise to rounding or truncating a decimal, so that digits that suggest a greater accuracy than the measurement itself can grant are removed. The remaining digits are called significant digits. For example, measurements with a ruler can seldom be made without a margin of error of at least 0.001 meters. If the sides of a rectangle are measured as 1.23 meters and 4.56 meters, then multiplication gives an area for the rectangle between 5.614591 square meters and 5.603011 square meters. Since not even the second digit after the decimal place is preserved, the following digits are fully insignificant. Therefore, sensibly, the result is usually rounded to 5.61. It can be shown that any ordered field, which is also complete, is isomorphic to the real numbers. The real numbers are not, however, an algebraically closed field, because they do not include a solution (often addressed as the square root of minus one) to the algebraic equation $x^2 + 1 = 0$.

Complex Numbers

Moving to a greater level of abstraction, the real numbers can be extended to the complex numbers. This set of numbers arose historically from trying to find closed formulas for the roots of cubic and quartic polynomials. This led to expressions involving the square roots of negative numbers, and eventually to the definition of a new number: a square root of −1, denoted by i, a symbol assigned by Leonhard Euler, and called the imaginary unit. The complex numbers consist of all numbers of the form

$$a + bi$$

where a and b are real numbers. Because of this, complex numbers correspond to points on the complex plane, a vector space of two real dimensions. In the expression a + bi, the real number a is called the real part and b is called the imaginary part. If the real part of a complex number is 0, then the number is called an imaginary number or is referred to as purely imaginary; if the imaginary part is 0, then the number is a real number. Thus the real numbers are a subset of the complex numbers. If the real and imaginary parts of a complex number are both integers, then the number is called a Gaussian integer. The symbol for the complex numbers is **C** or $\mathbb{C}$.

In abstract algebra, the complex numbers are an example of an algebraically closed field, meaning that every polynomial with complex coefficients can be factored into linear factors. Like the real number system, the complex number system is a field and is complete, but unlike the real numbers, it is not ordered. That is, there is no consistent meaning assignable to saying that i is greater than 1, nor is there any meaning in saying that i is less than 1. In technical terms, the complex numbers lack of a total order that is compatible with field operations.

Subclasses of the Integers

Even and Odd Numbers

An even number is an integer that is "evenly divisible" by two, that is divisible by two without remainder; an odd number is an integer that is not even. (The old-fashioned term "evenly divisible" is now almost always shortened to "divisible".) Equivalently, another way of defining an odd number is that it is an integer of the form $n = 2k + 1$, where k is an integer, and an even number has the form $n = 2k$ where k is an integer.

Prime Numbers

A prime number is an integer greater than 1 that is not the product of two smaller positive integers. The first few prime numbers are 2, 3, 5, 7, and 11. The prime numbers have been widely studied for more than 2000 years and have led to many questions, only some of which have been answered. The study of these questions is called number theory. An example of a question that is still unanswered is whether every even number is the sum of two primes. This is called Goldbach's conjecture.

A question that has been answered is whether every integer greater than one is a product of primes in only one way, except for a rearrangement of the primes. This is called fundamental theorem of arithmetic. A proof appears in Euclid's Elements.

Other classes of Integers

Many subsets of the natural numbers have been the subject of specific studies and have been named, often after the first mathematician that has studied them. Example of such sets of integers are Fibonacci numbers and perfect numbers. For more examples, see Integer sequence.

Subclasses of the Complex Numbers

Algebraic, Irrational and Transcendental Numbers

Algebraic numbers are those that are a solution to a polynomial equation with integer coefficients. Real numbers that are not rational numbers are called irrational numbers. Complex numbers which are not algebraic are called transcendental numbers. The algebraic numbers that are solutions of a monic polynomial equation with integer coefficients are called algebraic integers.

Computable Numbers

A computable number, also known as recursive number, is a real number such that there exists an algorithm which, given a positive number n as input, produces the first n digits of the computable number's decimal representation. Equivalent definitions can be given using μ-recursive functions, Turing machines or λ-calculus. The computable numbers are stable for all usual arithmetic operations, including the computation of the roots of a polynomial, and thus form a real closed field that contains the real algebraic numbers.

The computable numbers may be viewed as the real numbers that may be exactly represented in a computer: a computable number is exactly represented by its first digits and a program for computing further digits. However, the computable numbers are rarely used in practice. One reason is that there is no algorithm for testing the equality of two computable numbers. More precisely, there cannot exist any algorithm which takes any computable number as an input, and decides in every case if this number is equal to zero or not.

The set of computable numbers has the same cardinality as the natural numbers. Therefore, almost all real numbers are non-computable. However, it is very difficult to produce explicitly a real number that is not computable.

Extensions of the Concept

p-adic Numbers

The p-adic numbers may have infinitely long expansions to the left of the decimal point, in the same way that real numbers may have infinitely long expansions to the right. The number system that results depends on what base is used for the digits: any base is possible, but a prime number base provides the best mathematical properties. The set of the p-adic numbers contains the rational numbers, but is not contained in the complex numbers.

The elements of an algebraic function field over a finite field and algebraic numbers have many similar properties. Therefore, they are often regarded as numbers by number theorists. The p-adic numbers play an important role in this analogy.

Hypercomplex Numbers

Some number systems that are not included in the complex numbers may be constructed from the real numbers in a way that generalize the construction of the complex numbers. They are sometimes called hypercomplex numbers. They include the quaternions **H**, introduced by Sir William Rowan Hamilton, in which multiplication is not commutative, and the octonions, in which multiplication is not associative.

Transfinite Numbers

For dealing with infinite sets, the natural numbers have been generalized to the ordinal numbers and to the cardinal numbers. The former gives the ordering of the set, while the latter gives its size. For finite sets, both ordinal and cardinal numbers are identified with the natural numbers. In the infinite case, many ordinal numbers correspond to the same cardinal number.

Nonstandard Numbers

Hyperreal numbers are used in non-standard analysis. The hyperreals, or nonstandard reals (usually denoted as ***R**), denote an ordered field that is a proper extension of the ordered field of real numbers **R** and satisfies the transfer principle. This principle allows true first-order statements about **R** to be reinterpreted as true first-order statements about ***R**.

Superreal and surreal numbers extend the real numbers by adding infinitesimally small numbers and infinitely large numbers, but still form fields.

A relation number is defined as the class of relations consisting of all those relations that are similar to one member of the class.

HISTORY

First Use of Numbers

Bones and other artifacts have been discovered with marks cut into them that many believe are tally marks. These tally marks may have been used for counting elapsed time, such as numbers of days, lunar cycles or keeping records of quantities, such as of animals.

A tallying system has no concept of place value (as in modern decimal notation), which limits its representation of large numbers. Nonetheless tallying systems are considered the first kind of abstract numeral system.

The first known system with place value was the Mesopotamian base 60 system (ca. 3400 BC) and the earliest known base 10 system dates to 3100 BC in Egypt.

Zero

The use of 0 as a number should be distinguished from its use as a placeholder numeral in place-value systems. Many ancient texts used 0. Babylonian (Modern Iraq) and Egyptian texts used it. Egyptians used the word nfr to denote zero balance in double entry accounting entries. Indian texts used a Sanskrit word Shunye or shunya to refer to the concept of void. In mathematics texts this word often refers to the number zero.

The number 605 in Khmer numerals, from an inscription from 683 AD. An early use of zero as a decimal figure.

Records show that the Ancient Greeks seemed unsure about the status of 0 as a number: they asked themselves "how can 'nothing' be something?" leading to interesting philosophical and, by the Medieval period, religious arguments about the nature and existence of 0 and the vacuum. The paradoxes of Zeno of Elea depend in large part on the uncertain interpretation of 0. (The ancient Greeks even questioned whether 1 was a number.)

The late Olmec people of south-central Mexico began to use a true zero (a shell glyph) in the New World possibly by the 4th century BC but certainly by 40 BC, which became an integral part of Maya numerals and the Maya calendar. Mayan arithmetic used base 4 and base 5 written as base 20. Sanchez in 1961 reported a base 4, base 5 "finger" abacus.

By 130 AD, Ptolemy, influenced by Hipparchus and the Babylonians, was using a symbol for 0 (a small circle with a long overbar) within a sexagesimal numeral system otherwise using alphabetic Greek numerals. Because it was used alone, not as just a placeholder, this Hellenistic zero was the first documented use of a true zero in the Old World. In later Byzantine manuscripts of his Syntaxis Mathematica (Almagest), the Hellenistic zero had morphed into the Greek letter omicron (otherwise meaning 70).

Another true zero was used in tables alongside Roman numerals by 525 (first known use by Dionysius Exiguus), but as a word, nulla meaning nothing, not as a symbol. When division produced 0 as a remainder, nihil, also meaning nothing, was used. These medieval zeros were used by all future medieval computists (calculators of Easter). An isolated use of their initial, N, was used in a table of Roman numerals by Bede or a colleague about 725, a true zero symbol.

An early documented use of the zero by Brahmagupta (in the Brāhmasphu□asiddhānta) dates to 628. He treated 0 as a number and discussed operations involving it, including division. By this time (the 7th century) the concept had clearly reached Cambodia as Khmer numerals, and documentation shows the idea later spreading to China and the Islamic world.

Negative Numbers

The abstract concept of negative numbers was recognized as early as 100 BC – 50 BC in China. The Nine Chapters on the Mathematical Art contains methods for finding the areas of figures; red rods were used to denote positive coefficients, black for negative. The first reference in a Western work was in the 3rd century AD in Greece. Diophantus referred to the equation equivalent to $4x + 20 = 0$ (the solution is negative) in Arithmetica, saying that the equation gave an absurd result.

During the 600s, negative numbers were in use in India to represent debts. Diophantus' previous reference was discussed more explicitly by Indian mathematician Brahmagupta, in Brāhmasphu□asiddhānta 628, who used negative numbers to produce the general form quadratic formula that remains in use today. However, in the 12th century in India, Bhaskara gives negative roots for quadratic equations but says the negative value "is in this case not to be taken, for it is inadequate; people do not approve of negative roots."

European mathematicians, for the most part, resisted the concept of negative numbers until the 17th century, although Fibonacci allowed negative solutions in financial problems where they could be interpreted as debts (chapter 13 of Liber Abaci, 1202) and later as losses (in Flos). At the same time, the Chinese were indicating negative numbers by drawing a diagonal stroke through the right-most non-zero digit of the corresponding positive number's numeral. The first use of negative numbers in a European work was by Nicolas Chuquet during the 15th century. He used them as exponents, but referred to them as "absurd numbers".

As recently as the 18th century, it was common practice to ignore any negative results returned by equations on the assumption that they were meaningless, just as René Descartes did with negative solutions in a Cartesian coordinate system.

Rational Numbers

It is likely that the concept of fractional numbers dates to prehistoric times. The Ancient Egyptians used their Egyptian fraction notation for rational numbers in mathematical texts such as the Rhind Mathematical Papyrus and the Kahun Papyrus. Classical Greek and Indian mathematicians made studies of the theory of rational numbers, as part of the general study of number theory. The best known of these is Euclid's Elements, dating to roughly 300 BC. Of the Indian texts, the most relevant is the Sthananga Sutra, which also covers number theory as part of a general study of mathematics.

The concept of decimal fractions is closely linked with decimal place-value notation; the two seem to have developed in tandem. For example, it is common for the Jain math sutra to include calculations of decimal-fraction approximations to pi or the square root of 2. Similarly, Babylonian math texts had always used sexagesimal (base 60) fractions with great frequency.

Irrational Numbers

The earliest known use of irrational numbers was in the Indian Sulba Sutras composed between 800 and 500 BC. The first existence proofs of irrational numbers is usually attributed to Pythagoras, more specifically to the Pythagorean Hippasus of Metapontum, who produced a (most likely geometrical) proof of the irrationality of the square root of 2. The story goes that Hippasus discovered irrational numbers when trying to represent the square root of 2 as a fraction. However Pythagoras believed in the absoluteness of numbers, and could not accept the existence of irrational numbers. He could not disprove their existence through logic, but he could not accept irrational numbers, so he sentenced Hippasus to death by drowning.

The 16th century brought final European acceptance of negative integral and fractional numbers. By the 17th century, mathematicians generally used decimal fractions with modern notation. It was not, however, until the 19th century that mathematicians separated irrationals into algebraic and transcendental parts, and once more undertook scientific study of irrationals. It had remained almost dormant since Euclid. In 1872, the publication of the theories of Karl Weierstrass (by his pupil Kossak), Heine (Crelle, 74), Georg Cantor (Annalen, 5), and Richard Dedekind was brought about. In 1869, Méray had taken the same point of departure as Heine, but the theory is generally referred to the year 1872. Weierstrass's method was completely set forth by Salvatore Pincherle (1880), and Dedekind's has received additional prominence through the author's later work (1888) and endorsement by Paul Tannery (1894). Weierstrass, Cantor, and Heine base their theories on infinite series, while Dedekind founds his on the idea of a cut (Schnitt) in the system of real numbers, separating all rational numbers into two groups having certain characteristic properties. The subject has received later contributions at the hands of Weierstrass, Kronecker (Crelle, 101), and Méray.

The search for roots of quintic and higher degree equations was an important development, the Abel–Ruffini theorem (Ruffini 1799, Abel 1824) showed that they could not be solved by radicals (formulas involving only arithmetical operations and roots). Hence it was necessary to consider the wider set of algebraic numbers (all solutions to polynomial equations). Galois (1832) linked polynomial equations to group theory giving rise to the field of Galois theory.

Continued fractions, closely related to irrational numbers (and due to Cataldi, 1613), received attention at the hands of Euler, and at the opening of the 19th century were brought into prominence through the writings of Joseph Louis Lagrange. Other noteworthy contributions have been made by Druckenmüller (1837), Kunze (1857), Lemke (1870), and Günther (1872). Ramus (1855) first connected the subject with determinants, resulting, with the subsequent contributions of Heine, Möbius, and Günther, in the theory of Kettenbruchdeterminanten.

Transcendental Numbers and Reals

The existence of transcendental numbers was first established by Liouville (1844, 1851). Hermite proved in 1873 that e is transcendental and Lindemann proved in 1882 that π is transcendental. Finally, Cantor showed that the set of all real numbers is uncountably infinite but the set of all algebraic numbers is countably infinite, so there is an uncountably infinite number of transcendental numbers.

Infinity and Infinitesimals

The earliest known conception of mathematical infinity appears in the Yajur Veda, an ancient Indian script, which at one point states, "If you remove a part from infinity or add a part to infinity, still what remains is infinity." Infinity was a popular topic of philosophical study among the Jain mathematicians c. 400 BC. They distinguished between five types of infinity: infinite in one and two directions, infinite in area, infinite everywhere, and infinite perpetually.

Aristotle defined the traditional Western notion of mathematical infinity. He distinguished between actual infinity and potential infinity—the general consensus being that only the latter had true value. Galileo Galilei's Two New Sciences discussed the idea of one-to-one correspondences between infinite sets. But the next major advance in the theory was made by Georg Cantor; in 1895 he published a book about his new set theory, introducing, among other things, transfinite numbers and formulating the continuum hypothesis.

In the 1960s, Abraham Robinson showed how infinitely large and infinitesimal numbers can be rigorously defined and used to develop the field of nonstandard analysis. The system of hyperreal numbers represents a rigorous method of treating the ideas about infinite and infinitesimal numbers that had been used casually by mathematicians, scientists, and engineers ever since the invention of infinitesimal calculus by Newton and Leibniz.

A modern geometrical version of infinity is given by projective geometry, which introduces "ideal points at infinity", one for each spatial direction. Each family of parallel lines in a given direction is postulated to converge to the corresponding ideal point. This is closely related to the idea of vanishing points in perspective drawing.

Complex Numbers

The earliest fleeting reference to square roots of negative numbers occurred in the work of the mathematician and inventor Heron of Alexandria in the 1st century AD, when he considered the volume of an impossible frustum of a pyramid. They became more prominent when in the 16th century closed formulas for the roots of third and fourth degree polynomials were discovered by Italian mathematicians such as Niccolò Fontana Tartaglia and Gerolamo Cardano. It was soon realized that these formulas, even if one was only interested in real solutions, sometimes required the manipulation of square roots of negative numbers.

This was doubly unsettling since they did not even consider negative numbers to be on firm ground at the time. When René Descartes coined the term "imaginary" for these quantities in 1637, he intended it as derogatory. A further source of confusion was that the equation

$$\left(\sqrt{-1}\right)^2 = \sqrt{-1}\sqrt{-1} = -1$$

seemed capriciously inconsistent with the algebraic identity

$$\sqrt{a}\sqrt{b} = \sqrt{ab},$$

which is valid for positive real numbers a and b, and was also used in complex number calculations with one of a, b positive and the other negative. The incorrect use of this identity, and the related identity

$$\frac{1}{\sqrt{a}} = \sqrt{\frac{1}{a}}$$

in the case when both a and b are negative even bedeviled Euler. This difficulty eventually led him to the convention of using the special symbol i in place of $\sqrt{-1}$ to guard against this mistake.

The 18th century saw the work of Abraham de Moivre and Leonhard Euler. De Moivre's formula (1730) states:

$$(\cos\theta + i\sin\theta)^n = \cos n\theta + i\sin n\theta$$

and to Euler (1748) Euler's formula of complex analysis:

$$\cos\theta + i\sin\theta = e^{i\theta}.$$

The existence of complex numbers was not completely accepted until Caspar Wessel described the geometrical interpretation in 1799. Carl Friedrich Gauss rediscovered and popularized it several years later, and as a result the theory of complex numbers received a notable expansion. The idea of the graphic representation of complex numbers had appeared, however, as early as 1685, in Wallis's De Algebra tractatus.

Also in 1799, Gauss provided the first generally accepted proof of the fundamental theorem of algebra, showing that every polynomial over the complex numbers has a full set of solutions in that realm. The general acceptance of the theory of complex numbers is due to the labors of Augustin Louis Cauchy and Niels Henrik Abel, and especially the latter, who was the first to boldly use complex numbers with a success that is well known.

Gauss studied complex numbers of the form a + bi, where a and b are integral, or rational (and i is one of the two roots of $x^2 + 1 = 0$). His student, Gotthold Eisenstein, studied the type a + bω, where ω is a complex root of $x^3 - 1 = 0$. Other such classes (called cyclotomic fields) of complex numbers derive from the roots of unity $x^k - 1 = 0$ for higher values of k. This generalization is largely due to Ernst Kummer, who also invented ideal numbers, which were expressed as geometrical entities by Felix Klein in 1893.

In 1850 Victor Alexandre Puiseux took the key step of distinguishing between poles and branch points, and introduced the concept of essential singular points. This eventually led to the concept of the extended complex plane.

Prime Numbers

Prime numbers have been studied throughout recorded history. Euclid devoted one book of the Elements to the theory of primes; in it he proved the infinitude of the primes and the fundamental theorem of arithmetic, and presented the Euclidean algorithm for finding the greatest common divisor of two numbers.

In 240 BC, Eratosthenes used the Sieve of Eratosthenes to quickly isolate prime numbers. But further development of the theory of primes in Europe dates to the Renaissance and later eras.

In 1796, Adrien-Marie Legendre conjectured the prime number theorem, describing the asymptotic distribution of primes. Other results concerning the distribution of the primes include Euler's proof that the sum of the reciprocals of the primes diverges, and the Goldbach conjecture, which claims that any sufficiently large even number is the sum of two primes. Yet another conjecture related to the distribution of prime numbers is the Riemann hypothesis, formulated by Bernhard Riemann in 1859. The prime number theorem was finally proved by Jacques Hadamard and Charles de la Vallée-Poussin in 1896. Goldbach and Riemann's conjectures remain unproven and unrefuted.

NATURAL NUMBER

In mathematics, the natural numbers are those used for counting (as in "there are six coins on the table") and ordering (as in "this is the third largest city in the country"). In common language, words used for counting are "cardinal numbers" and words used for ordering are "ordinal numbers".

Some definitions, including the standard ISO 80000-2, begin the natural numbers with 0, corresponding to the non-negative integers0, 1, 2, 3, ..., whereas others start with 1, corresponding to the positive integers 1, 2, 3, ... Texts that exclude zero from the natural numbers sometimes refer to the natural numbers together with zero as the whole numbers, but in other writings, that term is used instead for the integers (including negative integers).

The natural numbers are the basis from which many other number sets may be built by extension: the integers, by including an additive inverse (−n) for each natural number n (and zero, if it is not there already, as its own additive inverse); the rational numbers, by including a multiplicative inverse (1/n) for each nonzero integer n; the real numbers by including with the rationals the (converging) Cauchy sequences of rationals; the complex numbers, by including with the real numbers the unresolved square root of minus one; and so on. These chains of extensions make the natural numbers canonically embedded (identified) in the other number systems.

Properties of the natural numbers, such as divisibility and the distribution of prime numbers, are studied in number theory. Problems concerning counting and ordering, such as partitioning and enumerations, are studied in combinatorics.

Natural numbers can be used for counting (one apple, two apples, three apples, ...).

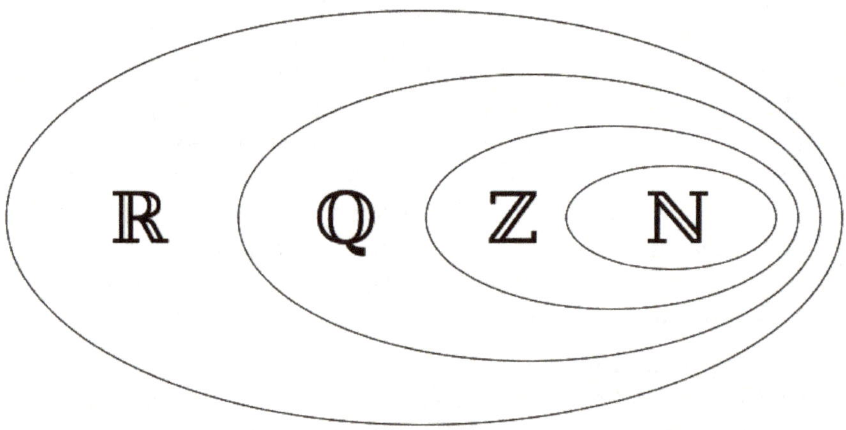

Real numbers (R) include the rational (Q), which include the integers (Z), which include the natural numbers (N).

In common language, for example in primary school, natural numbers may be called counting numbers to contrast the discreteness of counting to the continuity of measurement, established by the real numbers.

The natural numbers can, at times, appear as a convenient set of names (labels), that is, as what linguists call nominal numbers, foregoing many or all of the properties of being a number in a mathematical sense.

History

Ancient Roots

The most primitive method of representing a natural number is to put down a mark for each object. Later, a set of objects could be tested for equality, excess or shortage, by striking out a mark and removing an object from the set.

The first major advance in abstraction was the use of numerals to represent numbers. This allowed systems to be developed for recording large numbers. The ancient Egyptians developed a powerful system of numerals with distinct hieroglyphs for 1, 10, and all the powers of 10 up to over 1 million. A stone carving from Karnak, dating from around 1500 BC and now at the Louvre in Paris, depicts 276 as 2 hundreds, 7 tens, and 6 ones; and similarly for the number 4,622. The Babylonians had a place-value system based essentially on the numerals for 1 and 10, using base sixty, so that the symbol for sixty was the same as the symbol for one, its value being determined from context.

The Ishango bone (on exhibition at the Royal Belgian Institute of Natural Sciences) is believed to have been used 20,000 years ago for natural number arithmetic.

A much later advance was the development of the idea that 0 can be considered as a number, with its own numeral. The use of a 0 digit in place-value notation (within other numbers) dates back as early as 700 BC by the Babylonians, but they omitted such a digit when it would have been the last symbol in the number. The Olmec and Maya civilizations used 0 as a separate number as early as the 1st century BC, but this usage did not spread beyond Mesoamerica. The use of a numeral 0 in modern times originated with the Indian mathematician Brahmagupta in 628. However, 0 had been used as a number in the medieval computus (the calculation of the date of Easter), beginning with Dionysius Exiguus in 525, without being denoted by a numeral (standard Roman numerals do not have a symbol for 0); instead nulla (or the genitive form nullae) from nullus, the Latin word for "none", was employed to denote a 0 value.

The first systematic study of numbers as abstractions is usually credited to the Greek philosophers Pythagoras and Archimedes. Some Greek mathematicians treated the number 1 differently than larger numbers, sometimes even not as a number at all.

Independent studies also occurred at around the same time in India, China, and Mesoamerica.

Modern Definitions

In 19th century Europe, there was mathematical and philosophical discussion about the exact nature of the natural numbers. A school of Naturalism stated that the natural numbers were a direct consequence of the human psyche. Henri Poincaré was one of its advocates, as was Leopold Kronecker who summarized "God made the integers, all else is the work of man".

In opposition to the Naturalists, the constructivists saw a need to improve the logical rigor in the foundations of mathematics. In the 1860s, Hermann Grassmann suggested a recursive definition for natural numbers thus stating they were not really natural but a consequence of definitions. Later, two classes of such formal definitions were constructed; later, they were shown to be equivalent in most practical applications.

Set-theoretical definitions of natural numbers were initiated by Frege and he initially defined a natural number as the class of all sets that are in one-to-one correspondence with a particular set, but this definition turned out to lead to paradoxes including Russell's paradox. Therefore, this formalism was modified so that a natural number is defined as a particular set, and any set that can be put into one-to-one correspondence with that set is said to have that number of elements.

The second class of definitions was introduced by Giuseppe Peano and is now called Peano arithmetic. It is based on an axiomatization of the properties of ordinal numbers: each natural number has a successor and every non-zero natural number has a unique predecessor. Peano arithmetic is equiconsistent with several weak systems of set theory. One such system is ZFC with the axiom of infinity replaced by its negation. Theorems that can be proved in ZFC but cannot be proved using the Peano Axioms include Goodstein's theorem.

With all these definitions it is convenient to include 0 (corresponding to the empty set) as a natural number. Including 0 is now the common convention among set theorists and logicians. Other mathematicians also include 0 although many have kept the older tradition and take 1 to be the first natural number. Computer scientists often start from zero when enumerating items like loop counters and string- or array- elements.

Notation

Mathematicians use **N** or $\mathbb{N}$ (an N in blackboard bold) to refer to the set of all natural numbers. This set is countably infinite: it is infinite but countable by definition. This is also expressed by saying that the cardinal number of the set is aleph-naught $\aleph_0$.

The double-struck capital N symbol, often used to denote the set of all natural numbers.

To be unambiguous about whether 0 is included or not, sometimes an index (or superscript) "0" is added in the former case, and a superscript "*" or subscript ">0" is added in the latter case:

$\mathbb{N}^0 = \mathbb{N}_0 = \{0, 1, 2, \ldots\}$

$\mathbb{N}^* = \mathbb{N}^+ = \mathbb{N}_1 = \mathbb{N}_{>0} = \{1, 2, \ldots\}$.

Alternately natural numbers may be distinguished from positive integers with the notation, but it must be understood by context that, since both symbols are used, the natural number do contain zero.

$\mathbb{N} = \{0,1, 2, \ldots\}$.

$\mathbb{Z}^+ = \{1, 2, \ldots\}$.

Properties

Addition

One can recursively define an addition on the natural numbers by setting $a + 0 = a$ and $a + S(b) = S(a + b)$ for all a, b. Here S should be read as "successor". This turns the natural numbers ($\mathbb{N}$, +) into a commutative monoid with identity element 0, the so-called free object with one generator. This monoid satisfies the cancellation property and can be embedded in a group (in the mathematical sense of the word group). The smallest group containing the natural numbers is the integers.

If 1 is defined as S(0), then $b + 1 = b + S(0) = S(b + 0) = S(b)$. That is, b + 1 is simply the successor of b.

Multiplication

Analogously, given that addition has been defined, a multiplication × can be defined via $a \times 0 = 0$ and $a \times S(b) = (a \times b) + a$. This turns $(\mathbb{N}^*, \times)$ into a free commutative monoid with identity element 1; a generator set for this monoid is the set of prime numbers.

Relationship between Addition and Multiplication

Addition and multiplication are compatible, which is expressed in the distribution law: $a \times (b + c) = (a \times b) + (a \times c)$. These properties of addition and multiplication make the natural numbers an instance of a commutative semiring. Semirings are an algebraic generalization of the natural numbers where multiplication is not necessarily commutative. The lack of additive inverses, which is equivalent to the fact that $\mathbb{N}$ is not closed under subtraction, means that $\mathbb{N}$ is not a ring; instead it is a semiring (also known as a rig).

If the natural numbers are taken as "excluding 0", and "starting at 1", the definitions of + and × are as above, except that they begin with $a + 1 = S(a)$ and $a \times 1 = a$.

Order

In this section, juxtaposed variables such as ab indicate the product $a \times b$, and the standard order of operations is assumed.

A total order on the natural numbers is defined by letting $a \leq b$ if and only if there exists another natural number c with $a + c = b$. This order is compatible with the arithmetical operations in the following sense: if a, b and c are natural numbers and $a \leq b$, then $a + c \leq b + c$ and $ac \leq bc$. An important property of the natural numbers is that they are well-ordered: every non-empty set of natural numbers has a least element. The rank among well-ordered sets is expressed by an ordinal number; for the natural numbers this is expressed as ω.

Division

In this section, juxtaposed variables such as ab indicate the product $a \times b$, and the standard order of operations is assumed.

While it is in general not possible to divide one natural number by another and get a natural number as result, the procedure of division with remainder is available as a substitute: for any two natural numbers a and b with $b \neq 0$ there are natural numbers q and r such that

$a = bq + r$ and $r < b$.

The number q is called the quotient and r is called the remainder of division of a by b. The numbers q and r are uniquely determined by a and b. This Euclidean division is key to several other properties (divisibility), algorithms (such as the Euclidean algorithm), and ideas in number theory.

Algebraic Properties Satisfied by the Natural Numbers

The addition (+) and multiplication (×) operations on natural numbers as defined above have several algebraic properties:

- Closure under addition and multiplication: for all natural numbers a and b, both a + b and a × b are natural numbers.
- Associativity: for all natural numbers a, b, and c, a + (b + c) = (a + b) + c and a × (b × c) = (a × b) × c.
- Commutativity: for all natural numbers a and b, a + b = b + a and a × b = b × a.
- Existence of identity elements: for every natural number a, a + 0 = a and a × 1 = a.
- Distributivity of multiplication over addition for all natural numbers a, b, and c, a × (b + c) = (a × b) + (a × c).
- No nonzero zero divisors: if a and b are natural numbers such that a × b = 0, then a = 0 or b = 0.

Generalizations

Two generalizations of natural numbers arise from the two uses:

- A natural number can be used to express the size of a finite set; more generally a cardinal number is a measure for the size of a set also suitable for infinite sets; this refers to a concept of "size" such that if there is a bijection between two sets they have the same size. The set of natural numbers itself and any other countably infinite set has cardinality aleph-null ($\aleph_0$).
- Linguistic ordinal numbers "first", "second", "third" can be assigned to the elements of a totally ordered finite set, and also to the elements of well-ordered countably infinite sets like the set of natural numbers itself. This can be generalized to ordinal numbers which describe the position of an element in a well-ordered set in general. An ordinal number is also used to describe the "size" of a well-ordered set, in a sense different from cardinality: if there is an order isomorphism between two well-ordered sets they have the same ordinal number. The first ordinal number that is not a natural number is expressed as ω; this is also the ordinal number of the set of natural numbers itself.

Many well-ordered sets with cardinal number $\aleph_0$ have an ordinal number greater than ω (the latter is the lowest possible). The least ordinal of cardinality $\aleph_0$ (i.e., the initial ordinal) is ω.

For finite well-ordered sets, there is one-to-one correspondence between ordinal and cardinal numbers; therefore they can both be expressed by the same natural number, the number of elements of the set. This number can also be used to describe the position of an element in a larger finite, or an infinite, sequence.

A countable non-standard model of arithmetic satisfying the Peano Arithmetic (i.e., the first-order Peano axioms) was developed by Skolem in 1933. The hypernatural numbers are an uncountable model that can be constructed from the ordinary natural numbers via the ultrapower construction.

Georges Reeb used to claim provocatively that The naïve integers don't fill up ℕ. Other generalizations are discussed in the article on numbers.

Formal Definitions

Peano Axioms

Many properties of the natural numbers can be derived from the Peano axioms.

- Axiom One: 0 is a natural number.
- Axiom Two: Every natural number has a successor.
- Axiom Three: 0 is not the successor of any natural number.
- Axiom Four: If the successor of x equals the successor of y, then x equals y.
- Axiom Five (the axiom of induction): If a statement is true of 0, and if the truth of that statement for a number implies its truth for the successor of that number, then the statement is true for every natural number.

These are not the original axioms published by Peano, but are named in his honor. Some forms of the Peano axioms have 1 in place of 0. In ordinary arithmetic, the successor of x is x + 1. Replacing Axiom Five by an axiom schema one obtains a (weaker) first-order theory called Peano Arithmetic.

Constructions based on Set Theory

von Neumann Construction

In the area of mathematics called set theory, a special case of the von Neumann ordinal construction defines the natural numbers as follows:

- Set 0 = { }, the empty set,
- Define S(a) = a ∪ {a} for every set a. S(a) is the successor of a, and S is called the successor function.
- By the axiom of infinity, there exists a set which contains 0 and is closed under the successor function. Such sets are said to be 'inductive'. The intersection of all such inductive sets is defined to be the set of natural numbers. It can be checked that the set of natural numbers satisfies the Peano axioms.
- It follows that each natural number is equal to the set of all natural numbers less than it:
- 0 = { },
- 1 = 0 ∪ {0} = {0} = {{ }},
- 2 = 1 ∪ {1} = {0, 1} = {{ }, {{ }}},

- $3 = 2 \cup \{2\} = \{0, 1, 2\} = \{\{\ \}, \{\{\ \}\}, \{\{\ \}, \{\{\ \}\}\}\}$,
- $n = n-1 \cup \{n-1\} = \{0, 1, \ldots, n-1\} = \{\{\ \}, \{\{\ \}\}, \ldots, \{\{\ \}, \{\{\ \}\}, \ldots\}\}$, etc.

With this definition, a natural number n is a particular set with n elements, and $n \leq m$ if and only if n is a subset of m.

Also, with this definition, different possible interpretations of notations like $\mathbb{R}^n$ (n-tuples versus mappings of n into $\mathbb{R}$) coincide.

Even if one does not accept the axiom of infinity and therefore cannot accept that the set of all natural numbers exists, it is still possible to define any one of these sets.

Other Constructions

Although the standard construction is useful, it is not the only possible construction. Zermelo's construction goes as follows:

- Set $0 = \{\ \}$
- Define $S(a) = \{a\}$,
- It then follows that
- $0 = \{\ \}$,
- $1 = \{0\} = \{\{\ \}\}$,
- $2 = \{1\} = \{\{\{\ \}\}\}$,
- $n = \{n-1\} = \{\{\{\ldots\}\}\}$, etc.

Each natural number is then equal to the set of the natural number preceding it.

PRIME NUMBER

A prime number (or a prime) is a natural number greater than 1 that has no positive divisors other than 1 and itself. A natural number greater than 1 that is not a prime number is called a composite number. For example, 5 is prime because 1 and 5 are its only positive integer factors, whereas 6 is composite because it has the divisors 2 and 3 in addition to 1 and 6. The fundamental theorem of arithmetic establishes the central role of primes in number theory: any integer greater than 1 can be expressed as a product of primes that is unique up to ordering. The uniqueness in this theorem requires excluding 1 as a prime because one can include arbitrarily many instances of 1 in any factorization, e.g., 3, $1 \cdot 3$, $1 \cdot 1 \cdot 3$, etc. are all valid factorizations of 3.

The property of being prime (or not) is called primality. A simple but slow method of verifying the primality of a given number n is known as trial division. It consists of testing whether n is a multiple of any integer between 2 and $\sqrt{n}$. Algorithms much more efficient than trial division have been devised to test the primality of large numbers. These include the Miller–Rabin primality test, which is fast but has a small probability of error, and the AKS primality test, which always produces the correct answer in polynomial time but is too slow to be practical. Particularly fast methods are available for numbers of special forms, such as Mersenne numbers. As of January 2016, the largest known prime number has 22,338,618 decimal digits.

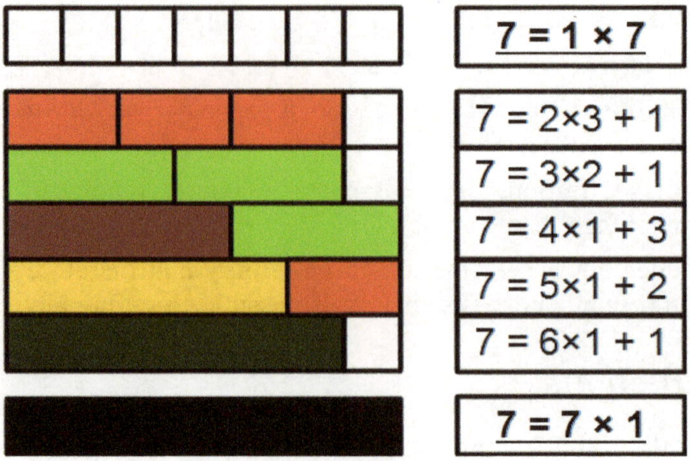

Demonstration, with Cuisenaire rods, that the number 7 is prime, being divisible only by 1 and 7.

There are infinitely many primes, as demonstrated by Euclid around 300 BC. There is no known simple formula that separates prime numbers from composite numbers. However, the distribution of primes, that is to say, the statistical behavior of primes in the large, can be modelled. The first result in that direction is the prime number theorem, proven at the end of the 19th century, which says that the probability that a given, randomly chosen number n is prime is inversely proportional to its number of digits, or to the logarithm of n.

Many questions regarding prime numbers remain open, such as Goldbach's conjecture (that every even integer greater than 2 can be expressed as the sum of two primes), and the twin prime conjecture (that there are infinitely many pairs of primes whose difference is 2). Such questions spurred the development of various branches of number theory, focusing on analytic or algebraic aspects of numbers. Primes are used in several routines in information technology, such as public-key cryptography, which makes use of properties such as the difficulty of factoring large numbers into their prime factors. Prime numbers give rise to various generalizations in other mathematical domains, mainly algebra, such as prime elements and prime ideals.

Definition and Examples

A natural number (i.e. 1, 2, 3, 4, 5, 6, etc.) is called a prime number (or a prime) if it has exactly two positive divisors, 1 and the number itself. Natural numbers greater than 1 that are not prime are called composite.

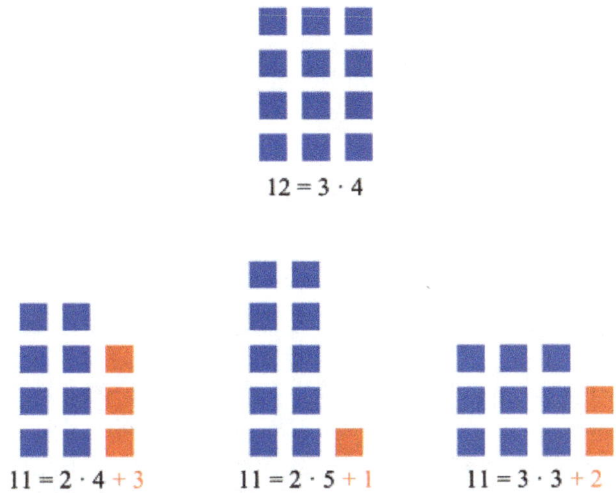

The number 12 is not a prime, as 12 items can be placed into 3 equal-size columns of 4 each (among other ways). 11 items cannot be all placed into several equal-size columns of more than 1 item each without some extra items leftover (a remainder). Therefore, the number 11 is a prime.

Among the numbers 1 to 6, the numbers 2, 3, and 5 are the prime numbers, while 1, 4, and 6 are not prime. 1 is excluded as a prime number, for reasons explained below. 2 is a prime number, since the only natural numbers dividing it are 1 and 2. Next, 3 is prime, too: 1 and 3 do divide 3 without remainder, but 3 divided by 2 gives remainder 1. Thus, 3 is prime. However, 4 is composite, since 2 is another number (in addition to 1 and 4) dividing 4 without remainder:

$4 = 2 \cdot 2.$

5 is again prime: none of the numbers 2, 3, or 4 divide 5. Next, 6 is divisible by 2 or 3, since

$6 = 2 \cdot 3.$

Hence, 6 is not prime. The image at the right illustrates that 12 is not prime: $12 = 3 \cdot 4$. No even number greater than 2 is prime because by definition, any such number n has at least three distinct divisors, namely 1, 2, and n. This implies that n is not prime. Accordingly, the term odd prime refers to any prime number greater than 2. Similarly, when written in the usual decimal system, all prime numbers larger than 5 end in 1, 3, 7, or 9, since even numbers are multiples of 2 and numbers ending in 0 or 5 are multiples of 5.

If n is a natural number, then 1 and n divide n without remainder. Therefore, the condition of being a prime can also be restated as: a number is prime if it is greater than one and if none of

$2, 3, \ldots, n - 1$

divides n (without remainder). Yet another way to say the same is: a number $n > 1$ is prime if it cannot be written as a product of two integers a and b, both of which are larger than 1:

$n = a \cdot b.$

In other words, n is prime if n items cannot be divided up into smaller equal-size groups of more than one item.

The set of all primes is often denoted by **P**.

The first 168 prime numbers (all the prime numbers less than 1000) are:

2, 3, 5, 7, 11, 13, 17, 19, 23, 29, 31, 37, 41, 43, 47, 53, 59, 61, 67, 71, 73, 79, 83, 89, 97, 101, 103, 107, 109, 113, 127, 131, 137, 139, 149, 151, 157, 163, 167, 173, 179, 181, 191, 193, 197, 199, 211, 223, 227, 229, 233, 239, 241, 251, 257, 263, 269, 271, 277, 281, 283, 293, 307, 311, 313, 317, 331, 337, 347, 349, 353, 359, 367, 373, 379, 383, 389, 397, 401, 409, 419, 421, 431, 433, 439, 443, 449, 457, 461, 463, 467, 479, 487, 491, 499, 503, 509, 521, 523, 541, 547, 557, 563, 569, 571, 577, 587, 593, 599, 601, 607, 613, 617, 619, 631, 641, 643, 647, 653, 659, 661, 673, 677, 683, 691, 701, 709, 719, 727, 733, 739, 743, 751, 757, 761, 769, 773, 787, 797, 809, 811, 821, 823, 827, 829, 839, 853, 857, 859, 863, 877, 881, 883, 887, 907, 911, 919, 929, 937, 941, 947, 953, 967, 971, 977, 983, 991, 997 (sequence A000040 in the OEIS).

Fundamental Theorem of Arithmetic

The crucial importance of prime numbers to number theory and mathematics in general stems from the fundamental theorem of arithmetic, which states that every integer larger than 1 can be written as a product of one or more primes in a way that is unique except for the order of the prime factors. Primes can thus be considered the "basic building blocks" of the natural numbers. For example:

23244	$= 2 \cdot 2 \cdot 3 \cdot 13 \cdot 149$
	$= 2^2 \cdot 3 \cdot 13 \cdot 149$. ($2^2$ denotes the square or second power of 2.)

As in this example, the same prime factor may occur multiple times. A decomposition:

$n = p_1 \cdot p_2 \cdot \ldots \cdot p_t$

of a number n into (finitely many) prime factors p_1, p_2, ... to p_t is called prime factorization of n. The fundamental theorem of arithmetic can be rephrased so as to say that any factorization into primes will be identical except for the order of the factors. So, albeit there are many prime factorization algorithms to do this in practice for larger numbers, they all have to yield the same result.

If p is a prime number and p divides a product ab of integers, then p divides a or p divides b. This proposition is known as Euclid's lemma. It is used in some proofs of the uniqueness of prime factorizations.

Primality of One

Most early Greeks did not even consider 1 to be a number, so they could not consider it to be a prime. By the Middle Ages and Renaissance many mathematicians included 1 as the first prime number. In the mid-18th century Christian Goldbach listed 1 as the first prime in his famous correspondence with Leonhard Euler; however, Euler himself did not consider 1 to be a prime number. In the 19th century many mathematicians still considered the number 1 to be a prime. For example, Derrick Norman Lehmer's list of primes up to 10,006,721, reprinted as late as 1956, started with 1 as its first prime. Henri Lebesgue is said to be the last professional mathematician to call 1 prime. By the early 20th century, mathematicians began to arrive at the consensus that 1 is not a prime number, but rather forms its own special category as a "unit".

A large body of mathematical work would still be valid when calling 1 a prime, but Euclid's fundamental theorem of arithmetic (mentioned above) would not hold as stated. For example, the number 15 can be factored as $3 \cdot 5$ and $1 \cdot 3 \cdot 5$; if 1 were admitted as a prime, these two presentations would be considered different factorizations of 15 into prime numbers, so the statement of that theorem would have to be modified. Similarly, the sieve of Eratosthenes would not work correctly if 1 were considered a prime: a modified version of the sieve that considers 1 as prime would eliminate all multiples of 1 (that is, all other numbers) and produce as output only the single number 1. Furthermore, the prime numbers have several properties that the number 1 lacks, such as the relationship of the number to its corresponding value of Euler's totient function or the sum of divisors function.

History

There are hints in the surviving records of the ancient Egyptians that they had some knowledge of prime numbers: the Egyptian fraction expansions in the Rhind papyrus, for instance, have quite different forms for primes and for composites.

1	2	3	4	5	6	7	8	9	10
11	12	13	14	15	16	17	18	19	20
21	22	23	24	25	26	27	28	29	30
31	32	33	34	35	36	37	38	39	40
41	42	43	44	45	46	47	48	49	50
51	52	53	54	55	56	57	58	59	60
61	62	63	64	65	66	67	68	69	70
71	72	73	74	75	76	77	78	79	80
81	82	83	84	85	86	87	88	89	90
91	92	93	94	95	96	97	98	99	100

The Sieve of Eratosthenes is a simple algorithm for finding all prime numbers up to a specified integer. It was created in the 3rd century BC by Eratosthenes, an ancient Greek mathematician.

However, the earliest surviving records of the explicit study of prime numbers come from the Ancient Greeks. Euclid's Elements (circa 300 BC) contain important theorems about primes, including the infinitude of primes and the fundamental theorem of arithmetic. Euclid also showed how to construct a perfect number from a Mersenne prime. The Sieve of Eratosthenes, attributed to Eratosthenes, is a simple method to compute primes, although the large primes found today with computers are not generated this way.

After the Greeks, little happened with the study of prime numbers until the 17th century. In 1640 Pierre de Fermat stated (without proof) Fermat's little theorem (later proved by Leibniz and Euler). Fermat also conjectured that all numbers of the form $2^{2n} + 1$ are prime (they are called Fermat numbers) and he verified this up to n = 4 (or $2^{16} + 1$). However, the very next Fermat number $2^{32} + 1$ is composite (one of its prime factors is 641), as Euler discovered later, and in fact no further Fermat numbers are known to be prime. The French monk Marin Mersenne looked at primes of the form $2^p - 1$, with p a prime. They are called Mersenne primes in his honor.

Euler's work in number theory included many results about primes. He showed the infinite series1/2 + 1/3 + 1/5 + 1/7 + 1/11 + ... is divergent. In 1747 he showed that the even perfect numbers are precisely the integers of the form $2^{p-1}(2^p - 1)$, where the second factor is a Mersenne prime.

At the start of the 19th century, Legendre and Gauss independently conjectured that as x tends to infinity, the number of primes up to x is asymptotic to x/ln(x), where ln(x) is the natural logarithm of x. Ideas of Riemann in his 1859 paper on the zeta-function sketched a program that would lead to a proof of the prime number theorem. This outline was completed by Hadamard and de la Vallée Poussin, who independently proved the prime number theorem in 1896.

Proving a number is prime is not done (for large numbers) by trial division. Many mathematicians have worked on primality tests for large numbers, often restricted to specific number forms. This includes Pépin's test for Fermat numbers (1877), Proth's theorem (around 1878), the Lucas–Lehmer primality test (originated 1856), and the generalized Lucas primality test. More recent algorithms like APRT-CL, ECPP, and AKS work on arbitrary numbers but remain much slower.

For a long time, prime numbers were thought to have extremely limited application outside of pure mathematics. This changed in the 1970s when the concepts of public-key cryptography were invented, in which prime numbers formed the basis of the first algorithms such as the RSA cryptosystem algorithm.

Since 1951 all the largest known primes have been found by computers. The search for ever larger primes has generated interest outside mathematical circles. The Great Internet Mersenne Prime Search and other distributed computing projects to find large primes have become popular, while mathematicians continue to struggle with the theory of primes.

Number of Prime Numbers

There are infinitely many prime numbers. Another way of saying this is that the sequence

2, 3, 5, 7, 11, 13, ...

of prime numbers never ends. This statement is referred to as Euclid's theorem in honor of the ancient Greek mathematician Euclid, since the first known proof for this statement is attributed to him. Many more proofs of the infinitude of primes are known, including an analytical proof by Euler, Goldbach's proof based on Fermat numbers, Furstenberg's proof using general topology, and Kummer's elegant proof.

Euclid's Proof

Euclid's proof (Book IX, Proposition 20) considers any finite set S of primes. The key idea is to consider the product of all these numbers plus one:

$$N = 1 + \prod_{p \in S} p.$$

Like any other natural number, N is divisible by at least one prime number (it is possible that N itself is prime).

None of the primes by which N is divisible can be members of the finite set S of primes with which we started, because dividing N by any one of these leaves a remainder of 1. Therefore, the primes by which N is divisible are additional primes beyond the ones we started with. Thus any finite set of primes can be extended to a larger finite set of primes.

It is often erroneously reported that Euclid begins with the assumption that the set initially considered contains all prime numbers, leading to a contradiction, or that it contains precisely the n smallest primes rather than any arbitrary finite set of primes. Today, the product of the smallest n primes plus 1 is conventionally called the nth Euclid number.

Euler's Analytical Proof

Euler's proof uses the partial sums of the reciprocals of primes,

$$S(p) = \frac{1}{2} + \frac{1}{3} + \frac{1}{5} + \frac{1}{7} + \cdots + \frac{1}{p}.$$

For any arbitrary real number x, there exists a prime p for which this partial sum is bigger than x. This shows that there are infinitely many primes, because if there were finitely many primes the sum would reach its maximum value at the biggest prime rather than being unbounded. More precisely, the growth rate of S(p) is doubly logarithmic, as quantified by Mertens' second theorem. For comparison, the sum

$$\frac{1}{1^2} + \frac{1}{2^2} + \frac{1}{3^2} + \cdots + \frac{1}{n^2} = \sum_{i=1}^{n} \frac{1}{i^2}$$

does not grow to infinity as n goes to infinity. In this sense, prime numbers occur more often than squares of natural numbers. Brun's theorem states that the sum of the reciprocals of twin primes,

$$\left(\frac{1}{3} + \frac{1}{5}\right) + \left(\frac{1}{5} + \frac{1}{7}\right) + \left(\frac{1}{11} + \frac{1}{13}\right) + \cdots = \sum_{\substack{p \text{ prime}, \\ p+2 \text{ prime}}} \left(\frac{1}{p} + \frac{1}{p+2}\right),$$

is finite. Because of Brun's theorem, it is not possible to use Euler's method to solve the twin prime conjecture, that there exist infinitely many twin primes.

Testing Primality and Integer Factorization

There are various methods to determine whether a given number n is prime. The most basic routine, trial division, is of little practical use because of its slowness. One group of modern primality tests is applicable to arbitrary numbers, while more efficient tests are available for particular numbers. Most such methods only tell whether n is prime or not. Routines also yielding one (or all) prime factors of n are called factorization algorithms.

Trial Division

The most basic method of checking the primality of a given integer n is called trial division. This routine consists of dividing n by each integer m that is greater than 1 and less than or equal to the square root of n. If the result of any of these divisions is an integer, then n is not a prime, otherwise it is a prime. Indeed, if n=ab is composite (with a and b $\neq$ 1) then one of the factors a or b is necessarily at most $\sqrt{n}$. For example, for n=37, the trial divisions are by m = 2, 3, 4, 5, and 6. None of these numbers divides 37, so 37 is prime. This routine can be implemented more efficiently if a complete list of primes up to $\sqrt{n}$ is known—then trial divisions need to be checked only for those m that are prime. For example, to check the primality of 37, only three divisions are necessary (m = 2, 3, and 5), given that 4 and 6 are composite.

While a simple method, trial division quickly becomes impractical for testing large integers because the number of possible factors grows too rapidly as n increases. According to the prime number theorem explained below, the number of prime numbers less than $\sqrt{n}$ is approximately given by $\sqrt{n}/\ln(\sqrt{n})$, so the algorithm may need up to this number of trial divisions to check the primality of n. For n = 10^{20}, this number is 450 million—too large for many practical applications.

Sieves

An algorithm yielding all primes up to a given limit, such as required in the primes-only trial division method, is called a prime number sieve. The oldest example, the sieve of Eratosthenes (see above), is still the most commonly used. The sieve of Atkin is another option. Before the advent of computers, lists of primes up to bounds like 10^7 were also used.

Primality Testing Versus Primality Proving

Modern primality tests for general numbers n can be divided into two main classes, probabilistic (or "Monte Carlo") and deterministic algorithms. Deterministic algorithms provide a way to tell for sure whether a given number is prime or not. For example, trial division is a deterministic algorithm because, if performed correctly, it will always identify a prime number as prime and a composite number as composite. Probabilistic algorithms are normally faster, but do not completely prove that a number is prime. These tests rely on testing a given number in a partly random way. For example, a given test might pass all the time if applied to a prime number, but pass only with probability p if applied to a composite number. If we repeat the test n times and pass every time, then the probability that our number is composite is $1/(1\text{-}p)^n$, which decreases exponentially with the number of tests, so we can be as sure as we like (though never perfectly sure) that the number is prime. On the other hand, if the test ever fails, then we know that the number is composite.

A particularly simple example of a probabilistic test is the Fermat primality test, which relies on the fact (Fermat's little theorem) that $n^p \equiv n \pmod p$ for any n if p is a prime number. If we have a number b that we want to test for primality, then we work out $n^b \pmod b$ for a random value of n as our test. A flaw with this test is that there are some composite numbers (the Carmichael numbers) that satisfy the Fermat identity even though they are not prime, so the test has no way of distinguishing between prime numbers and Carmichael numbers. Carmichael numbers are substantially rarer than prime numbers, though, so this test can be useful for practical purposes. More powerful extensions of the Fermat primality test, such as the Baillie-PSW, Miller-Rabin, and Solovay-Strassen tests, are guaranteed to fail at least some of the time when applied to a composite number.

Deterministic algorithms do not erroneously report composite numbers as prime. In practice, the fastest such method is known as elliptic curve primality proving. Analyzing its run time is based on heuristic arguments, as opposed to the rigorously proven complexity of the more recent AKS primality test. Deterministic methods are typically slower than probabilistic ones, so the latter ones are typically applied first before a more time-consuming deterministic routine is employed.

The following table lists a number of prime tests. The running time is given in terms of n, the number to be tested and, for probabilistic algorithms, the number k of tests performed. Moreover, ε is an arbitrarily small positive number, and log is the logarithm to an unspecified base. The big O notation means that, for example, elliptic curve primality proving requires a time that is bounded by a factor (not depending on n, but on ε) times $\log^{5+\varepsilon}(n)$.

Test	**Developed in**	**Type**	**Running time**	**Notes**
AKS primality test	2002	deterministic	$O(\log^{6+\varepsilon}(n))$	
Elliptic curve primality proving	1977	deterministic	$O(\log^{5+\varepsilon}(n))$ heuristically	
Baillie-PSW primality test	1980	probabilistic	$O(\log^3 n)$	no known counterexamples
Miller–Rabin primality test	1980	probabilistic	$O(k \cdot \log^{2+\varepsilon}(n))$	error probability 4^{-k}
Solovay–Strassen primality test	1977	probabilistic	$O(k \cdot \log^3 n)$	error probability 2^{-k}
Fermat primality test		probabilistic	$O(k \cdot \log^{2+\varepsilon}(n))$	fails for Carmichael numbers

Special-Purpose Algorithms and the Largest Known Prime

In addition to the aforementioned tests applying to any natural number n, a number of much more efficient primality tests is available for special numbers. For example, to run Lucas' primality test requires the knowledge of the prime factors of n − 1, while the Lucas–Lehmer primality test needs the prime factors of n + 1 as input. For example, these tests can be applied to check whether

$n! \pm 1 = 1 \cdot 2 \cdot 3 \cdot \ldots \cdot n \pm 1$

are prime. Prime numbers of this form are known as factorial primes. Other primes where either p + 1 or p − 1 is of a particular shape include the Sophie Germain primes (primes of the form 2p + 1 with p prime), primorial primes, Fermat primes and Mersenne primes, that is, prime numbers that are of the form $2^p - 1$, where p is an arbitrary prime. The Lucas–Lehmer test is particularly fast for numbers of this form. This is why the largest known prime has almost always been a Mersenne prime since the dawn of electronic computers.

The following table gives the largest known primes of the mentioned types. Some of these primes have been found using distributed computing. In 2009, the Great Internet Mersenne Prime Search project was awarded a US$100,000 prize for first discovering a prime with at least 10 million digits. The Electronic Frontier Foundation also offers $150,000 and $250,000 for primes with at least 100 million digits and 1 billion digits, respectively. Some of the largest primes not known to have any particular form (that is, no simple formula such as that of Mersenne primes) have been found by taking a piece of semi-random binary data, converting it to a number n, multiplying it by 256^k for some positive integer k, and searching for possible primes within the interval $[256^k n + 1, 256^k(n + 1) - 1]$.

Type	Prime	Number of decimal digits	Date	Found by
Mersenne prime	$2^{74,207,281} - 1$	22,338,618	January 7, 2016	Great Internet Mersenne Prime Search
not a Mersenne prime (Proth number)	$19,249 \times 2^{13,018,586} + 1$	3,918,990	March 26, 2007	Seventeen or Bust
factorial prime	150209! + 1	712,355	October 2011	PrimeGrid
primorial prime	1098133# - 1	476,311	March 2012	PrimeGrid
twin primes	$3756801695685 \times 2^{666669} \pm 1$	200,700	December 2011	PrimeGrid

Integer Factorization

Given a composite integer n, the task of providing one (or all) prime factors is referred to as factorization of n. Elliptic curve factorization is an algorithm relying on arithmetic on an elliptic curve.

Distribution

In 1975, number theorist Don Zagier commented that primes both

grow like weeds among the natural numbers, seeming to obey no other law than that of chance [but also] exhibit stunning regularity [and] that there are laws governing their behavior, and that they obey these laws with almost military precision.

The distribution of primes in the large, such as the question how many primes are smaller than a given, large threshold, is described by the prime number theorem, but no efficient formula for the n-th prime is known.

There are arbitrarily long sequences of consecutive non-primes, as for every positive integer n the n consecutive integers from $(n+1)!+2$ to $(n+1)!+n+1$ (inclusive) are all composite (as $(n+1)!+k$ is divisible by k for k between 2 and $n+1$).

Dirichlet's theorem on arithmetic progressions, in its basic form, asserts that linear polynomials

$$p(n) = a + bn$$

with coprime integers a and b take infinitely many prime values. Stronger forms of the theorem state that the sum of the reciprocals of these prime values diverges, and that different such polynomials with the same b have approximately the same proportions of primes.

The corresponding question for quadratic polynomials is less well-understood.

Formulas for Primes

There is no known efficient formula for primes. For example, Mills' theorem and a theorem of Wright assert that there are real constants A>1 and μ such that

$$\left\lfloor A^{3^n} \right\rfloor \text{ and } \left\lfloor 2^{\cdot^{\cdot^{\cdot^{2^{2^{\mu}}}}}} \right\rfloor$$

are prime for any natural number n. Here $\lfloor - \rfloor$ represents the floor function, i.e., largest integer not greater than the number in question. The latter formula can be shown using Bertrand's postulate (proven first by Chebyshev), which states that there always exists at least one prime number p with n < p < 2n − 2, for any natural number n > 3. However, computing A or μ requires the knowledge of infinitely many primes to begin with. Another formula is based on Wilson's theorem and generates the number 2 many times and all other primes exactly once.

There is no non-constant polynomial, even in several variables, that takes only prime values. However, there is a set of Diophantine equations in 9 variables and one parameter with the following property: the parameter is prime if and only if the resulting system of equations has a solution over the natural numbers. This can be used to obtain a single formula with the property that all its positive values are prime.

Number of Prime Numbers below a Given Number

The prime counting function π(n) is defined as the number of primes not greater than n.

For example, $\pi(11) = 5$, since there are five primes less than or equal to 11. There are known algorithms to compute exact values of $\pi(n)$ faster than it would be possible to compute each prime up to n. The prime number theorem states that $\pi(n)$ is approximately given by

$$\pi(n) \approx \frac{n}{\ln n},$$

in the sense that the ratio of $\pi(n)$ and the right hand fraction approaches 1 when n grows to infinity. This implies that the likelihood that a number less than n is prime is (approximately) inversely proportional to the number of digits in n. A more accurate estimate for $\pi(n)$ is given by the offset logarithmic integral

$$\mathrm{Li}(n) = \int_2^n \frac{dt}{\ln t}.$$

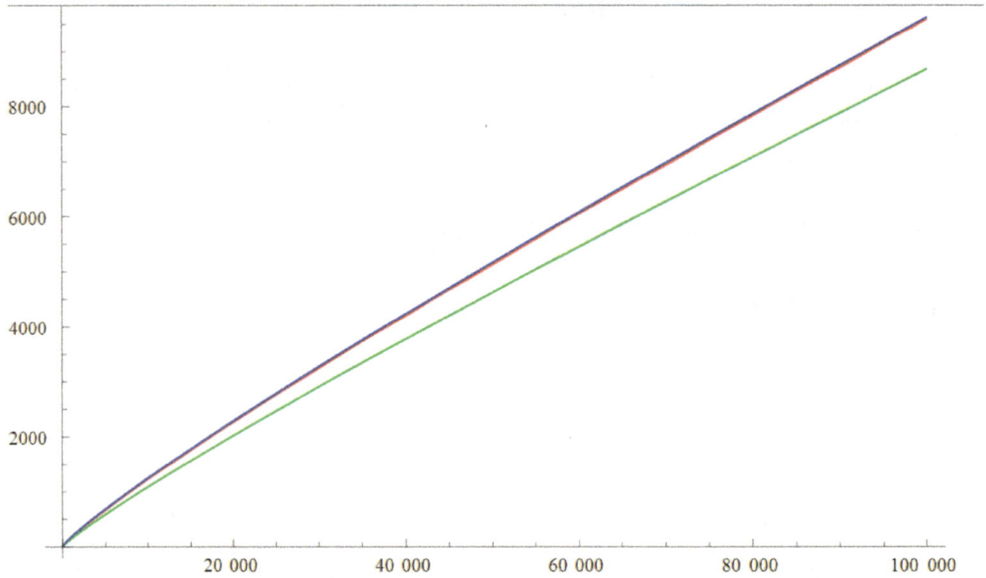

A chart depicting $\pi(n)$ (blue), n / ln (n) (green) and Li(n) (red)

The prime number theorem also implies estimates for the size of the n-th prime number p_n (i.e., $p_1 = 2$, $p_2 = 3$, etc.): up to a bounded factor, p_n grows like n log(n). In particular, the prime gaps, i.e. the differences $p_n - p_{n-1}$ of two consecutive primes, become arbitrarily large. This latter statement can also be seen in a more elementary way by noting that the sequence n! + 2, n! + 3, …, n! + n (for the notation n! read factorial) consists of n − 1 composite numbers, for any natural number n.

Arithmetic Progressions

An arithmetic progression is the set of natural numbers that give the same remainder when divided by some fixed number q called modulus. For example,

3, 12, 21, 30, 39, ...,

is an arithmetic progression modulo q = 9. Except for 3, none of these numbers is prime, since 3 + 9n = 3(1 + 3n) so that the remaining numbers in this progression are all composite. (In general terms, all prime numbers above q are of the form q#·n + m, where 0 < m < q#, and m has no prime factor ≤ q.) Thus, the progression

a, a + q, a + 2q, a + 3q, ...

can have infinitely many primes only when a and q are coprime, i.e., their greatest common divisor is one. If this necessary condition is satisfied, Dirichlet's theorem on arithmetic progressions asserts that the progression contains infinitely many primes. The picture below illustrates this with q = 9: the numbers are "wrapped around" as soon as a multiple of 9 is passed. Primes are highlighted in red. The rows (=progressions) starting with a = 3, 6, or 9 contain at most one prime number. In all other rows (a = 1, 2, 4, 5, 7, and 8) there are infinitely many prime numbers. What is more, the primes are distributed equally among those rows in the long run—the density of all primes congruent a modulo 9 is 1/6.

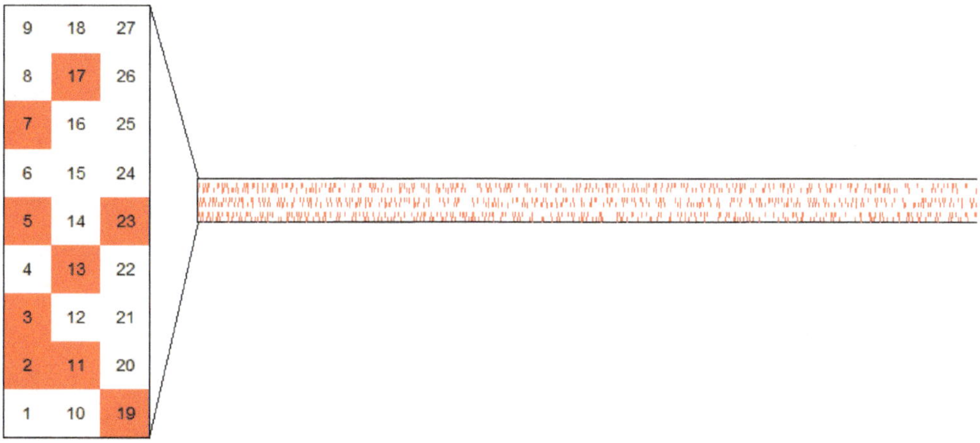

The Green–Tao theorem shows that there are arbitrarily long arithmetic progressions consisting of primes. An odd prime p is expressible as the sum of two squares, $p = x^2 + y^2$, exactly if p is congruent 1 modulo 4 (Fermat's theorem on sums of two squares).

Prime Values of Quadratic Polynomials

Euler noted that the function

$$n^2 + n + 41$$

yields prime numbers for $0 \leq n < 40$, a fact leading into deep algebraic number theory: more specifically, Heegner numbers. For greater n, the expression also produces composite values. The Hardy-Littlewood conjecture F makes an asymptotic prediction about the density of primes among the values of quadratic polynomials (with integer coefficients a, b, and c),

$$f(n) = ax^2 + bx + c,$$

in terms of Li(n) and the coefficients a, b, and c. However, progress has been difficult. No quadratic polynomial (with $a \neq 0$) is known to take infinitely many prime values. The Ulam spiral depicts all natural numbers in a spiral-like way. Primes cluster on certain diagonals and not others, suggesting that some quadratic polynomials take prime values more often than others.

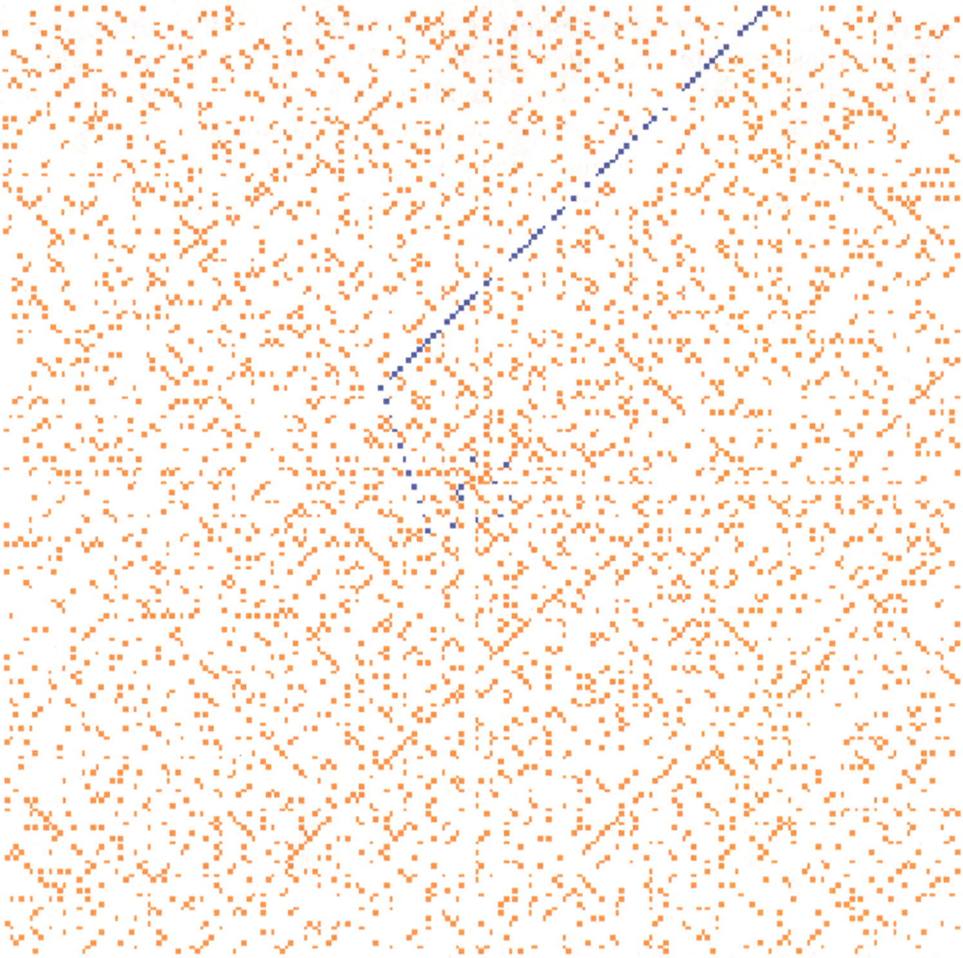

The Ulam spiral. Red pixels show prime numbers. Primes of the form $4n^2 - 2n + 41$ are highlighted in blue.

Zeta Function and the Riemann Hypothesis

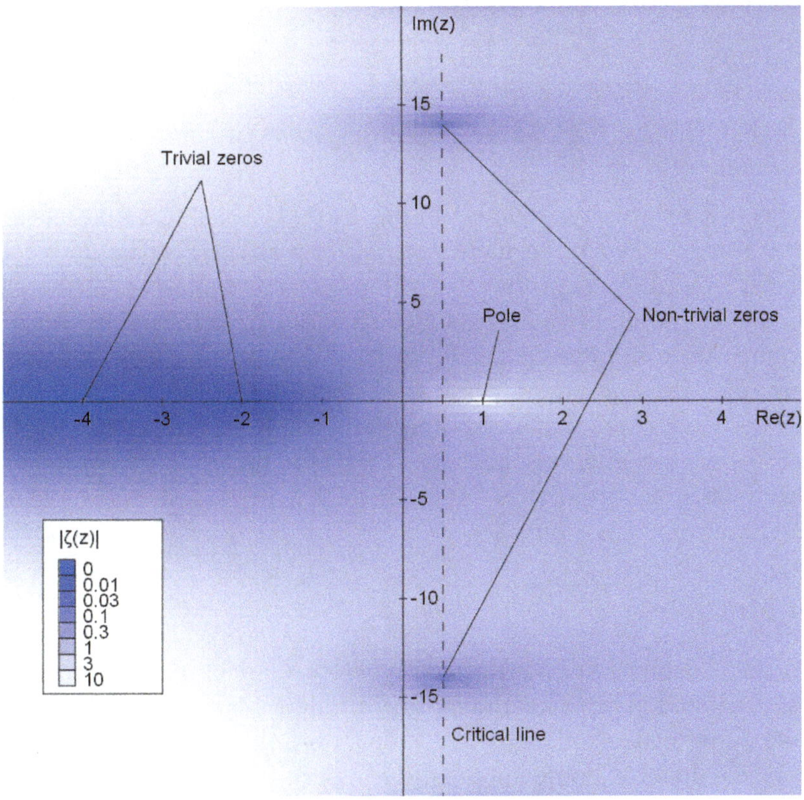

Plot of the zeta function ζ(s). At s=1, the function has a pole, that is to say, it tends to infinity.

The Riemann zeta function ζ(s) is defined as an infinite sum

$$\zeta(s) = \sum_{n=1}^{\infty} \frac{1}{n^s},$$

where s is a complex number with real part bigger than 1. It is a consequence of the fundamental theorem of arithmetic that this sum agrees with the infinite product

$$\prod_{p \text{ prime}} \frac{1}{1 - p^{-s}}.$$

The zeta function is closely related to prime numbers. For example, the aforementioned fact that there are infinitely many primes can also be seen using the zeta function: if there were only finitely many primes then ζ(1) would have a finite value. However, the harmonic series1 + 1/2 + 1/3 + 1/4 + ... diverges (i.e., exceeds any given number), so there must be infinitely many primes. Another example of the richness of the zeta function and a glimpse of modern algebraic number theory is the following identity (Basel problem), due to Euler,

$$\zeta(2) = \prod_p \frac{1}{1 - p^{-2}} = \frac{\pi^2}{6}.$$

The reciprocal of ζ(2), $6/\pi^2$, is the probability that two numbers selected at random are relatively prime.

The unproven Riemann hypothesis, dating from 1859, states that except for s = −2, −4, ..., all zeroes of the ζ-function have real part equal to 1/2. The connection to prime numbers is that it essentially says that the primes are as regularly distributed as possible. From a physical viewpoint, it roughly states that the irregularity in the distribution of primes only comes from random noise. From a mathematical viewpoint, it roughly states that the asymptotic distribution of primes (about x/log x of numbers less than x are primes, the prime number theorem) also holds for much shorter intervals of length about the square root of x (for intervals near x). This hypothesis is generally believed to be correct. In particular, the simplest assumption is that primes should have no significant irregularities without good reason.

Other Conjectures

In addition to the Riemann hypothesis, many more conjectures revolving about primes have been posed. Often having an elementary formulation, many of these conjectures have withstood a proof for decades: all four of Landau's problems from 1912 are still unsolved. One of them is Goldbach's conjecture, which asserts that every even integer n greater than 2 can be written as a sum of two primes. As of February 2011, this conjecture has been verified for all numbers up to $n = 2 \cdot 10^{17}$. Weaker statements than this have been proven, for example Vinogradov's theorem says that every sufficiently large odd integer can be written as a sum of three primes. Chen's theorem says that every sufficiently large even number can be expressed as the sum of a prime and a semiprime, the product of two primes. Also, any even integer can be written as the sum of six primes. The branch of number theory studying such questions is called additive number theory.

Other conjectures deal with the question whether an infinity of prime numbers subject to certain constraints exists. It is conjectured that there are infinitely many Fibonacci primes and infinitely many Mersenne primes, but not Fermat primes. It is not known whether or not there are an infinite number of Wieferich primes and of prime Euclid numbers.

A third type of conjectures concerns aspects of the distribution of primes. It is conjectured that there are infinitely many twin primes, pairs of primes with difference 2 (twin prime conjecture). Polignac's conjecture is a strengthening of that conjecture, it states that for every positive integer n, there are infinitely many pairs of consecutive primes that differ by 2n. It is conjectured there are infinitely many primes of the form $n^2 + 1$. These conjectures are special cases of the broad Schinzel's hypothesis H. Brocard's conjecture says that there are always at least four primes between the squares of consecutive primes greater than 2. Legendre's conjecture states that there is a prime number between n^2 and $(n + 1)^2$ for every positive integer n. It is implied by the stronger Cramér's conjecture.

Applications

For a long time, number theory in general, and the study of prime numbers in particular, was seen as the canonical example of pure mathematics, with no applications outside of the self-interest of studying the topic with the exception of use of prime numbered gear teeth to distribute wear evenly. In particular, number theorists such as British mathematician G. H. Hardy prided themselves on doing work that had absolutely no military significance. However, this vision was shattered in the 1970s, when it was publicly announced that prime numbers could be used as the basis for the creation of public key cryptography algorithms. Prime numbers are also used for hash tables and pseudorandom number generators.

Some rotor machines were designed with a different number of pins on each rotor, with the number of pins on any one rotor either prime, or coprime to the number of pins on any other rotor. This helped generate the full cycle of possible rotor positions before repeating any position.

The International Standard Book Numbers work with a check digit, which exploits the fact that 11 is a prime.

Arithmetic Modulo a Prime and Finite Fields

Modular arithmetic modifies usual arithmetic by only using the numbers

$$\{0, 1, 2, \dots, n-1\},$$

where n is a fixed natural number called modulus. Calculating sums, differences and products is done as usual, but whenever a negative number or a number greater than n − 1 occurs, it gets replaced by the remainder after division by n. For instance, for n = 7, the sum 3 + 5 is 1 instead of 8, since 8 divided by 7 has remainder 1. This is referred to by saying "3 + 5 is congruent to 1 modulo 7" and is denoted

$$3 + 5 \equiv 1 \pmod{7}.$$

Similarly, 6 + 1 ≡ 0 (mod 7), 2 − 5 ≡ 4 (mod 7), since −3 + 7 = 4, and 3 · 4 ≡ 5 (mod 7) as 12 has remainder 5. Standard properties of addition and multiplication familiar from the integers remain valid in modular arithmetic. In the parlance of abstract algebra, the above set of integers, which is also denoted **Z**/n**Z**, is therefore a commutative ring for any n. Division, however, is not in general possible in this setting. For example, for n = 6, the equation

$$3 \cdot x \equiv 2 \pmod{6},$$

a solution x of which would be an analogue of 2/3, cannot be solved, as one can see by calculating 3 · 0, ..., 3 · 5 modulo 6. The distinctive feature of prime numbers is the following: division is possible in modular arithmetic if and only if n is a prime. Equivalently, n is prime if and only if all integers m satisfying 2 ≤ m ≤ n − 1 are coprime to n, i.e. their only common divisor is one. Indeed, for n = 7, the equation

$$3 \cdot x \equiv 2 \pmod{7},$$

has a unique solution, x = 3. Because of this, for any prime p, **Z**/p**Z** (also denoted $\mathbf{F}_p$) is called a field or, more specifically, a finite field since it contains finitely many, namely p, elements.

A number of theorems can be derived from inspecting $\mathbf{F}_p$ in this abstract way. For example, Fermat's little theorem, stating

$$a^{p-1} \equiv 1 \pmod{p}$$

for any integer a not divisible by p, may be proved using these notions. This implies

$$\sum_{a=1}^{p-1} a^{p-1} \equiv (p-1) \cdot 1 \equiv -1 \pmod{p}.$$

Giuga's conjecture says that this equation is also a sufficient condition for p to be prime. Another consequence of Fermat's little theorem is the following: if p is a prime number other than 2 and 5, $^1/_p$ is always a recurring decimal, whose period is p − 1 or a divisor of p − 1. The fraction $^1/_p$ expressed likewise in base q (rather than base 10) has similar effect, provided that p is not a prime factor of q. Wilson's theorem says that an integer p > 1 is prime if and only if the factorial (p − 1)! + 1 is divisible by p. Moreover, an integer n > 4 is composite if and only if (n − 1)! is divisible by n.

Fermat Primes and Constructible Polygons

Fermat primes are primes of the form

$F_k = 2^{2k} + 1,$

with k a natural number. They are named after Pierre de Fermat, who conjectured that all such numbers are prime. This was based on the evidence of the first five numbers in this series—3, 5, 17, 257, and 65,537—being prime. However, F_5 is composite and so are all other Fermat numbers that have been verified as of 2015. A regular n-gon is constructible using straightedge and compass if and only if the odd prime factors of n (if any) are distinct Fermat primes.

Other Mathematical Occurrences of Primes

Many mathematical domains make great use of prime numbers. An example from the theory of finite groups are the Sylow theorems: if G is a finite group and p^n is the highest power of the prime p that divides the order of G, then G has a subgroup of order p^n. Also, any group of prime order is cyclic (Lagrange's theorem).

Public-Key Cryptography

Several public-key cryptography algorithms, such as RSA and the Diffie–Hellman key exchange, are based on large prime numbers (2048-bit primes are common). RSA relies on the assumption that it is much easier (i.e., more efficient) to perform the multiplication of two (large) numbers x and y than to calculate x and y (assumed coprime) if only the product xy is known. The Diffie–Hellman key exchange relies on the fact that there are efficient algorithms for modular exponentiation, while the reverse operation the discrete logarithm is thought to be a hard problem.

Prime Numbers in Nature

The evolutionary strategy used by cicadas of the genus Magicicada make use of prime numbers. These insects spend most of their lives as grubs underground. They only pupate and then emerge from their burrows after 7, 13 or 17 years, at which point they fly about, breed, and then die after a few weeks at most. The logic for this is believed to be that the prime number intervals between emergences make it very difficult for predators to evolve that could specialize as predators on Magicicadas. If Magicicadas appeared at a non-prime number intervals, say every 12 years, then predators appearing every 2, 3, 4, 6, or 12 years would be sure to meet them. Over a 200-year period, average predator populations during hypothetical outbreaks of 14- and 15-year cicadas would be up to 2% higher than during outbreaks of 13- and 17-year cicadas. Though small, this advantage appears to have been enough to drive natural selection in favour of a prime-numbered life-cycle for these insects.

There is speculation that the zeros of the zeta function are connected to the energy levels of complex quantum systems.

Generalizations

The concept of prime number is so important that it has been generalized in different ways in various branches of mathematics. Generally, "prime" indicates minimality or indecomposability, in an appropriate sense. For example, the prime field is the smallest subfield of a field F containing both 0 and 1. It is either **Q** or the finite field with p elements, whence the name. Often a second, additional meaning is intended by using the word prime, namely that any object can be, essentially uniquely, decomposed into its prime components. For example, in knot theory, a prime knot is a knot that is indecomposable in the sense that it cannot be written as the knot sum of two nontrivial knots. Any knot can be uniquely expressed as a connected sum of prime knots. Prime models and prime 3-manifolds are other examples of this type.

Prime Elements in Rings

Prime numbers give rise to two more general concepts that apply to elements of any commutative ring R, an algebraic structure where addition, subtraction and multiplication are defined: prime elements and irreducible elements. An element p of R is called prime element if it is neither zero nor a unit (i.e., does not have a multiplicative inverse) and satisfies the following requirement: given x and y in R such that p divides the product xy, then p divides x or y. An element is irreducible if it is not a unit and cannot be written as a product of two ring elements that are not units. In the ring **Z** of integers, the set of prime elements equals the set of irreducible elements, which is

$$\{\ldots, -11, -7, -5, -3, -2, 2, 3, 5, 7, 11, \ldots\}.$$

In any ring R, any prime element is irreducible. The converse does not hold in general, but does hold for unique factorization domains.

The fundamental theorem of arithmetic continues to hold in unique factorization domains. An example of such a domain is the Gaussian integers **Z**[i], that is, the set of complex numbers of the form a + bi where i denotes the imaginary unit and a and b are arbitrary integers. Its prime elements are known as Gaussian primes. Not every prime (in **Z**) is a Gaussian prime: in the bigger ring **Z**[i], 2 factors into the product of the two Gaussian primes (1 + i) and (1 − i). Rational primes (i.e. prime elements in **Z**) of the form 4k + 3 are Gaussian primes, whereas rational primes of the form 4k + 1 are not.

Prime Ideals

In ring theory, the notion of number is generally replaced with that of ideal. Prime ideals, which generalize prime elements in the sense that the principal ideal generated by a prime element is a prime ideal, are an important tool and object of study in commutative algebra, algebraic number theory and algebraic geometry. The prime ideals of the ring of integers are the ideals (0), (2), (3), (5), (7), (11), ... The fundamental theorem of arithmetic generalizes to the Lasker–Noether theorem, which expresses every ideal in a Noetherian commutative ring as an intersection of primary ideals, which are the appropriate generalizations of prime powers.

Prime ideals are the points of algebro-geometric objects, via the notion of the spectrum of a ring. Arithmetic geometry also benefits from this notion, and many concepts exist in both geometry and number theory. For example, factorization or ramification of prime ideals when lifted to an extension field, a basic problem of algebraic number theory, bears some resemblance with ramification in geometry. Such ramification questions occur even in number-theoretic questions solely concerned with integers. For example, prime ideals in the ring of integers of quadratic number fields can be used in proving quadratic reciprocity, a statement that concerns the solvability of quadratic equations

$$x^2 \equiv p \pmod{q},$$

where x is an integer and p and q are (usual) prime numbers. Early attempts to prove Fermat's Last Theorem climaxed when Kummer introduced regular primes, primes satisfying a certain requirement concerning the failure of unique factorization in the ring consisting of expressions

$$a_0 + a_1\zeta + \cdots + a_{p-1}\zeta^{p-1},$$

where a_0, ..., a_{p-1} are integers and ζ is a complex number such that $\zeta^p = 1$.

Valuations

Valuation theory studies certain functions from a field K to the real numbers **R** called valuations. Every such valuation yields a topology on K, and two valuations are called equivalent if they yield the same topology. A prime of K (sometimes called a place of K) is an equivalence class of valuations. For example, the p-adic valuation of a rational number q is defined to be the integer $v_p(q)$, such that

$$q = p^{v_p(q)} \frac{r}{s},$$

where both r and s are not divisible by p. For example, $v_3(18/7) = 2$. The p-adic norm is defined as[nb 1]

$$|q|_p := p^{-v_p(q)}.$$

In particular, this norm gets smaller when a number is multiplied by p, in sharp contrast to the usual absolute value (also referred to as the infinite prime). While completing **Q** (roughly, filling the gaps) with respect to the absolute value yields the field of real numbers, completing with respect to the p-adic norm $|-|_p$ yields the field of p-adic numbers. These are essentially all possible ways to complete **Q**, by Ostrowski's theorem. Certain arithmetic questions related to **Q** or more general global fields may be transferred back and forth to the completed (or local) fields. This local-global principle again underlines the importance of primes to number theory.

In the arts and literature

Prime numbers have influenced many artists and writers. The French composer Olivier Messiaen used prime numbers to create ametrical music through "natural phenomena". In works such as La Nativité du Seigneur (1935) and Quatre études de rythme (1949–50), he simultaneously employs motifs with lengths given by different prime numbers to create unpredictable rhythms: the primes 41, 43, 47 and 53 appear in the third étude, «Neumes rythmiques». According to Messiaen this way of composing was «inspired by the movements of nature, movements of free and unequal durations».

In his science fiction novel Contact, NASA scientist Carl Sagan suggested that prime numbers could be used as a means of communicating with aliens, an idea that he had first developed informally with American astronomer Frank Drake in 1975. In the novel The Curious Incident of the Dog in the Night-Time by Mark Haddon, the narrator arranges the sections of the story by consecutive prime numbers.

Many films, such as Cube, Sneakers, The Mirror Has Two Faces and A Beautiful Mind reflect a popular fascination with the mysteries of prime numbers and cryptography. Prime numbers are used as a metaphor for loneliness and isolation in the Paolo Giordano novel The Solitude of Prime Numbers, in which they are portrayed as "outsiders" among integers.

RATIONAL NUMBER

In mathematics, a rational number is any number that can be expressed as the quotient or fraction p/q of two integers, a numerator p and a non-zero denominator q. Since q may be equal to 1, every integer is a rational number. The set of all rational numbers, often referred to as "the rationals", is usually denoted by a boldface Q (or blackboard bold $\mathbb{Q}$, Unicode $\mathbb{Q}$); it was thus denoted in 1895 by Giuseppe Peano after quoziente, Italian for "quotient".

The decimal expansion of a rational number always either terminates after a finite number of digits or begins to repeat the same finite sequence of digits over and over. Moreover, any repeating or terminating decimal represents a rational number. These statements hold true not just for base 10, but also for any other integer base (e.g. binary, hexadecimal).

A real number that is not rational is called irrational. Irrational numbers include $\sqrt{2}$, π, e, and φ. The decimal expansion of an irrational number continues without repeating. Since the set of rational numbers is countable, and the set of real numbers is uncountable, almost all real numbers are irrational.

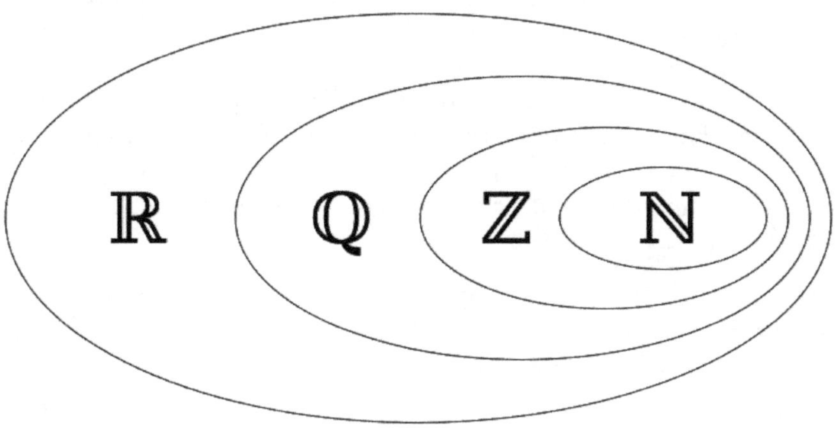

The rational numbers ($\mathbb{Q}$) are included in the real numbers, and in turn include the integers ($\mathbb{Z}$), which include the natural numbers ($\mathbb{N}$).

The rational numbers can be formally defined as the equivalence classes of the quotient set $(\mathbf{Z} \times (\mathbf{Z} \setminus \{0\})) / \sim$, where the cartesian product $\mathbf{Z} \times (\mathbf{Z} \setminus \{0\})$ is the set of all ordered pairs (m,n) where m and n are integers, n is not 0 ($n \neq 0$), and "$\sim$" is the equivalence relation defined by $(m_1,n_1) \sim (m_2,n_2)$ if, and only if, $m_1n_2 - m_2n_1 = 0$.

In abstract algebra, the rational numbers together with certain operations of addition and multiplication form the archetypical field of characteristic zero. As such, it is characterized as having no proper subfield or, alternatively, being the field of fractions for the ring of integers. Finite extensions of **Q** are called algebraic number fields, and the algebraic closure of **Q** is the field of algebraic numbers.

In mathematical analysis, the rational numbers form a dense subset of the real numbers. The real numbers can be constructed from the rational numbers by completion, using Cauchy sequences, Dedekind cuts, or infinite decimals.

Zero divided by any other integer equals zero; therefore, zero is a rational number (but division by zero is undefined).

Terminology

The term rational in reference to the set **Q** refers to the fact that a rational number represents a ratio of two integers. In mathematics, "rational" is often used as a noun abbreviating "rational number". The adjective rational sometimes means that the coefficients are rational numbers. For example, a rational point is a point with rational coordinates (that is a point whose coordinates are rational numbers; a rational matrix is a matrix of rational numbers; a rational polynomial may be a polynomial with rational coefficients, although the term "polynomial over the rationals" is generally preferred, for avoiding confusion with "rational expression" and "rational function" (a polynomial is a rational expression and defines a rational function, even if its coefficients are not rational numbers). However, a rational curve is not a curve defined over the rationals, but a curve which can be parameterized by rational functions.

Arithmetic

Embedding of Integers

Any integer n can be expressed as the rational number n/1.

Equality

$\frac{a}{b} = \frac{c}{d}$ if and only if $ad = bc$.

Ordering

Where both denominators are positive:

$\frac{a}{b} < \frac{c}{d}$ if and only if $ad < bc$.

If either denominator is negative, the fractions must first be converted into equivalent forms with positive denominators, through the equations:

$$\frac{-a}{-b} = \frac{a}{b}$$

and

$$\frac{a}{-b} = \frac{-a}{b}.$$

Addition

Two fractions are added as follows:

$$\frac{a}{b}+\frac{c}{d}=\frac{ad+bc}{bd}.$$

Subtraction

$$\frac{a}{b}-\frac{c}{d}=\frac{ad-bc}{bd}.$$

Multiplication

The rule for multiplication is:

$$\frac{a}{b}\cdot\frac{c}{d}=\frac{ac}{bd}.$$

Division

Where c ≠ 0:

$$\frac{a}{b}\div\frac{c}{d}=\frac{ad}{bc}.$$

Note that division is equivalent to multiplying by the reciprocal of the divisor fraction:

$$\frac{ad}{bc}=\frac{a}{b}\times\frac{d}{c}.$$

Inverse

Additive and multiplicative inverses exist in the rational numbers:

$$-\left(\frac{a}{b}\right)=\frac{-a}{b}=\frac{a}{-b}\quad\text{and}\quad\left(\frac{a}{b}\right)^{-1}=\frac{b}{a}\text{ if } a\neq 0.$$

Exponentiation to Integer Power

If n is a non-negative integer, then

$$\left(\frac{a}{b}\right)^n = \frac{a^n}{b^n}$$

and (if $a \neq 0$):

$$\left(\frac{a}{b}\right)^{-n} = \frac{b^n}{a^n}.$$

Continued Fraction Representation

A finite continued fraction is an expression such as

$$a_0 + \cfrac{1}{a_1 + \cfrac{1}{a_2 + \cfrac{1}{\ddots + \cfrac{1}{a_n}}}},$$

where a_n are integers. Every rational number a/b has two closely related expressions as a finite continued fraction, whose coefficients a_n can be determined by applying the Euclidean algorithm to (a,b).

Other Representations

- common fraction: $\frac{8}{3}$
- mixed numeral: $2\frac{2}{3}$
- repeating decimal using a vinculum: $2.\overline{6}$
- repeating decimal using parentheses: $2.(6)$
- continued fraction using traditional typography: $2 + \cfrac{1}{1 + \cfrac{1}{2}}$
- continued fraction in abbreviated notation: [2; 1, 2]
- egyptian fraction: $2 + \frac{1}{2} + \frac{1}{6}$
- prime power decomposition: $\frac{117}{1000} = 2^{-3} \times 3^2 \times 5^{-3} \times 13$
- quote notation: 3!6

are different ways to represent the same rational value.

Formal Construction

Mathematically we may construct the rational numbers as equivalence classes of ordered pairs of integers (m,n), with $n \neq 0$. This space of equivalence classes is the quotient space $(\mathbf{Z} \times (\mathbf{Z} \setminus \{0\})) / \sim$, where $(m_1,n_1) \sim (m_2,n_2)$ if, and only if, $m_1n_2 - m_2n_1 = 0$. We can define addition and multiplication of these pairs with the following rules:

$$(m_1, n_1) + (m_2, n_2) \equiv (m_1 n_2 + n_1 m_2, n_1 n_2)$$
$$(m_1, n_1) \times (m_2, n_2) \equiv (m_1 m_2, n_1 n_2)$$

and, if $m_2 \neq 0$, division by

$$\frac{(m_1, n_1)}{(m_2, n_2)} \equiv (m_1 n_2, n_1 m_2).$$

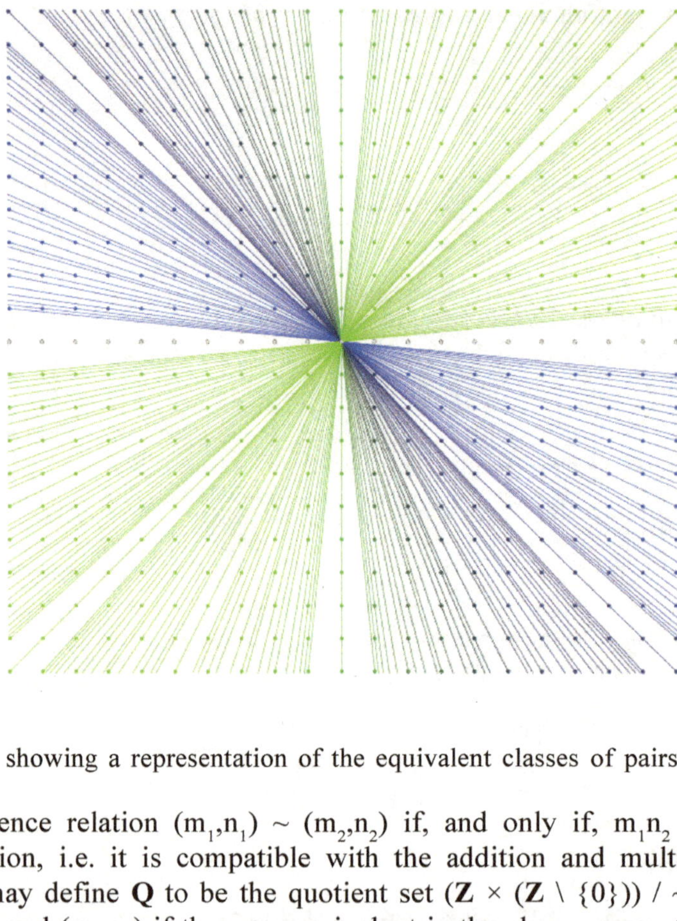

A diagram showing a representation of the equivalent classes of pairs of integers

The equivalence relation $(m_1,n_1) \sim (m_2,n_2)$ if, and only if, $m_1 n_2 - m_2 n_1 = 0$ is a congruence relation, i.e. it is compatible with the addition and multiplication defined above, and we may define **Q** to be the quotient set $(\mathbf{Z} \times (\mathbf{Z} \setminus \{0\})) / \sim$, i.e. we identify two pairs (m_1,n_1) and (m_2,n_2) if they are equivalent in the above sense. (This construction can be carried out in any integral domain: see field of fractions.) We denote by $[(m_1,n_1)]$ the equivalence class containing (m_1,n_1). If $(m_1,n_1) \sim (m_2,n_2)$ then, by definition, (m_1,n_1) belongs to $[(m_2,n_2)]$ and (m_2,n_2) belongs to $[(m_1,n_1)]$; in this case we can write $[(m_1,n_1)] = [(m_2,n_2)]$. Given any equivalence class $[(m,n)]$ there are a countably infinite number of representation, since

$$\cdots = [(-2m, -2n)] = [(-m, -n)] = [(m, n)] = [(2m, 2n)] = \cdots.$$

The canonical choice for $[(m,n)]$ is chosen so that n is positive and $\gcd(m,n) = 1$, i.e. m and n share no common factors, i.e. m and n are coprime. For example, we would write [(1,2)] instead of [(2,4)] or [(−12,−24)], even though [(1,2)] = [(2,4)] = [(−12,−24)].

We can also define a total order on **Q**. Let ∧ be the and-symbol and ∨ be the or-symbol. We say that $[(m_1,n_1)] \leq [(m_2,n_2)]$ if:

$$(n_1 n_2 > 0 \ \wedge \ m_1 n_2 \leq n_1 m_2) \ \vee \ (n_1 n_2 < 0 \ \wedge \ m_1 n_2 \geq n_1 m_2).$$

The integers may be considered to be rational numbers by the embedding that maps m to [(m,1)].

Properties

The set **Q**, together with the addition and multiplication operations shown above, forms a field, the field of fractions of the integers **Z**.

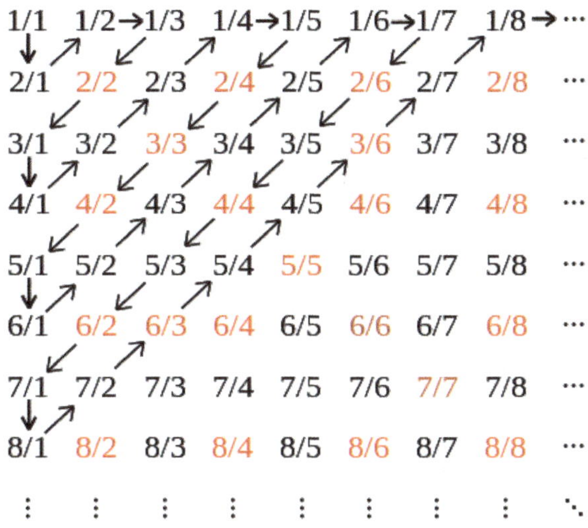

A diagram illustrating the countability of the rationals

The rationals are the smallest field with characteristic zero: every other field of characteristic zero contains a copy of **Q**. The rational numbers are therefore the prime field for characteristic zero.

The algebraic closure of **Q**, i.e. the field of roots of rational polynomials, is the algebraic numbers.

The set of all rational numbers is countable. Since the set of all real numbers is uncountable, we say that almost all real numbers are irrational, in the sense of Lebesgue measure, i.e. the set of rational numbers is a null set.

The rationals are a densely ordered set: between any two rationals, there sits another one, and, therefore, infinitely many other ones. For example, for any two fractions such that

$$\frac{a}{b} < \frac{c}{d}$$

(where b, d are positive), we have

$$\frac{a}{b} < \frac{ad+bc}{2bd} < \frac{c}{d}.$$

Any totally ordered set which is countable, dense (in the above sense), and has no least or greatest element is order isomorphic to the rational numbers.

Real Numbers and Topological Properties

The rationals are a dense subset of the real numbers: every real number has rational numbers arbitrarily close to it. A related property is that rational numbers are the only numbers with finite expansions as regular continued fractions.

By virtue of their order, the rationals carry an order topology. The rational numbers, as a subspace of the real numbers, also carry a subspace topology. The rational numbers form a metric space by using the absolute difference metric $d(x,y) = |x - y|$, and this yields a third topology on **Q**. All three topologies coincide and turn the rationals into a topological field. The rational numbers are an important example of a space which is not locally compact. The rationals are characterized topologically as the unique countable metrizable space without isolated points. The space is also totally disconnected. The rational numbers do not form a complete metric space; the real numbers are the completion of **Q** under the metric $d(x,y) = |x - y|$, above.

p-adic Numbers

In addition to the absolute value metric mentioned above, there are other metrics which turn **Q** into a topological field:

Let p be a prime number and for any non-zero integer a, let $|a|_p = p^{-n}$, where p^n is the highest power of p dividing a.

In addition set $|0|_p = 0$. For any rational number a/b, we set $|a/b|_p = |a|_p / |b|_p$.

Then $d_p(x,y) = |x - y|_p$ defines a metric on **Q**.

The metric space $(\mathbf{Q},d_p)$ is not complete, and its completion is the p-adic number field $\mathbf{Q}_p$. Ostrowski's theorem states that any non-trivial absolute value on the rational numbers **Q** is equivalent to either the usual real absolute value or a p-adic absolute value.

IRRATIONAL NUMBER

In mathematics, an irrational number is a real number that cannot be expressed as a ratio of integers, i.e. as a fraction. Therefore, irrational numbers, when written as decimal numbers, do not terminate, nor do they repeat. For example, the number π starts with 3.14159265358979, but no finite number of digits can represent it exactly and it does not end in a segment that repeats itself infinitely often. The same can be said for any irrational number.

As a consequence of Cantor's proof that the real numbers are uncountable and the rationals countable, it follows that almost all real numbers are irrational.

When the ratio of lengths of two line segments is irrational, the line segments are also described as being incommensurable, meaning they share no measure in common.

Numbers that are irrational include the ratio π of a circle's circumference to its diameter, Euler's number e, the golden ratio φ, and the square root of two; in fact all square roots of natural numbers, other than of perfect squares, are irrational.

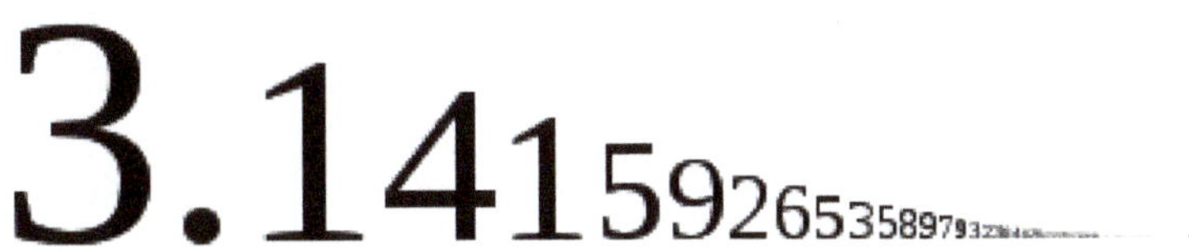

The mathematical constant pi (π) is an irrational number that is much represented in popular culture.

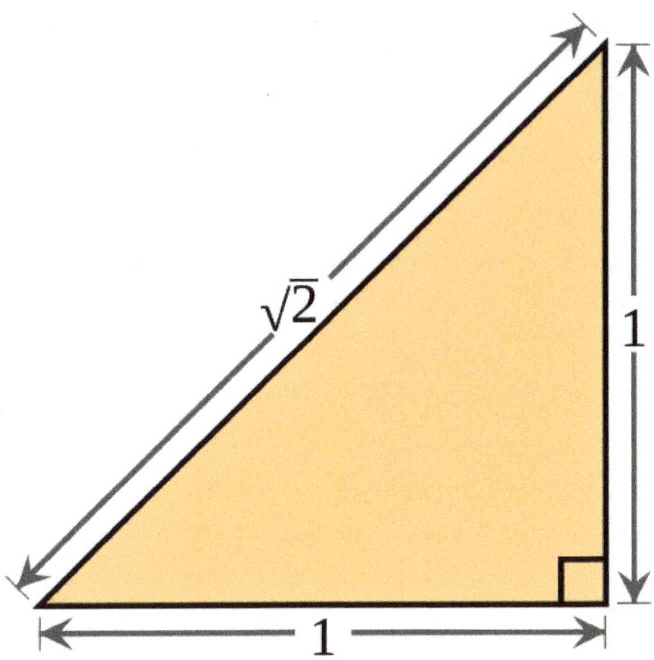

The number $\sqrt{2}$ is irrational.

History

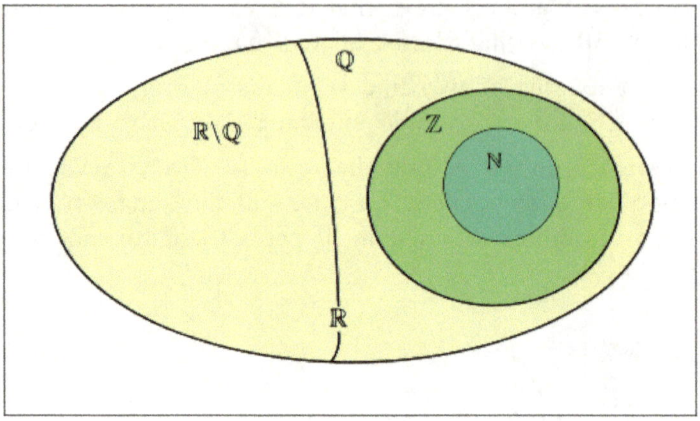

Set of real numbers (R), which include the rational (Q), which include the integers (Z), which include the natural numbers (N). The real numbers also include the irrational (R\Q).

Ancient Greece

The first proof of the existence of irrational numbers is usually attributed to a Pythagorean (possibly Hippasus of Metapontum), who probably discovered them while identifying sides of the pentagram. The then-current Pythagorean method would have claimed that there must be some sufficiently small, indivisible unit that could fit evenly into one of these lengths as well as the other. However, Hippasus, in the 5th century BC, was able to deduce that there was in fact no common unit of measure, and that the assertion of such an existence was in fact a contradiction. He did this by demonstrating that if the hypotenuse of an isosceles right triangle was indeed commensurable with a leg, then one of those lengths measured in that unit of measure must be both odd and even, which is impossible. His reasoning is as follows:

- Start with an isosceles right triangle with side lengths of integers a, b, and c. The ratio of the hypotenuse to a leg is represented by c:b.
- Assume a, b, and c are in the smallest possible terms (i.e. they have no common factors).
- By the Pythagorean theorem: $c^2 = a^2+b^2 = b^2+b^2 = 2b^2$. (Since the triangle is isosceles, a = b).
- Since $c^2 = 2b^2$, c^2 is divisible by 2, and therefore even.
- Since c^2 is even, c must be even.
- Since c and b have no common factors, and c is even, b must be odd (if b were even, b and c would have a common factor of 2).
- Since c is even, dividing c by 2 yields an integer. Let y be this integer ($c = 2y$).
- Squaring both sides of $c = 2y$ yields $c^2 = (2y)^2$, or $c^2 = 4y^2$.

- Substituting $4y^2$ for c^2 in the first equation ($c^2 = 2b^2$) gives us $4y^2 = 2b^2$.
- Dividing by 2 yields $2y^2 = b^2$.
- Since y is an integer, and $2y^2 = b^2$, b^2 is divisible by 2, and therefore even.
- Since b^2 is even, b must be even.
- However, we have already asserted that b must be odd, and b cannot be both odd and even. This contradiction proves that c and b cannot both be integers, and thus the existence of a number that cannot be expressed as a ratio of two integers.

Greek mathematicians termed this ratio of incommensurable magnitudes alogos, or inexpressible. Hippasus, however, was not lauded for his efforts: according to one legend, he made his discovery while out at sea, and was subsequently thrown overboard by his fellow Pythagoreans "...for having produced an element in the universe which denied the...doctrine that all phenomena in the universe can be reduced to whole numbers and their ratios." Another legend states that Hippasus was merely exiled for this revelation. Whatever the consequence to Hippasus himself, his discovery posed a very serious problem to Pythagorean mathematics, since it shattered the assumption that number and geometry were inseparable–a foundation of their theory.

The discovery of incommensurable ratios was indicative of another problem facing the Greeks: the relation of the discrete to the continuous. Brought into light by Zeno of Elea, who questioned the conception that quantities are discrete and composed of a finite number of units of a given size. Past Greek conceptions dictated that they necessarily must be, for "whole numbers represent discrete objects, and a commensurable ratio represents a relation between two collections of discrete objects." However Zeno found that in fact "[quantities] in general are not discrete collections of units; this is why ratios of incommensurable [quantities] appear....[Q]uantities are, in other words, continuous." What this means is that, contrary to the popular conception of the time, there cannot be an indivisible, smallest unit of measure for any quantity. That in fact, these divisions of quantity must necessarily be infinite. For example, consider a line segment: this segment can be split in half, that half split in half, the half of the half in half, and so on. This process can continue infinitely, for there is always another half to be split. The more times the segment is halved, the closer the unit of measure comes to zero, but it never reaches exactly zero. This is just what Zeno sought to prove. He sought to prove this by formulating four paradoxes, which demonstrated the contradictions inherent in the mathematical thought of the time. While Zeno's paradoxes accurately demonstrated the deficiencies of current mathematical conceptions, they were not regarded as proof of the alternative. In the minds of the Greeks, disproving the validity of one view did not necessarily prove the validity of another, and therefore further investigation had to occur.

The next step was taken by Eudoxus of Cnidus, who formalized a new theory of proportion that took into account commensurable as well as incommensurable quantities. Central to his idea was the distinction between magnitude and number. A magnitude "... was not a number but stood for entities such as line segments, angles, areas, volumes, and time which could vary, as we would say, continuously. Magnitudes were opposed to numbers, which jumped from one value to another, as from 4 to 5." Numbers are composed of some smallest, indivisible unit, whereas magnitudes are infinitely reducible.

Because no quantitative values were assigned to magnitudes, Eudoxus was then able to account for both commensurable and incommensurable ratios by defining a ratio in terms of its magnitude, and proportion as an equality between two ratios. By taking quantitative values (numbers) out of the equation, he avoided the trap of having to express an irrational number as a number. "Eudoxus' theory enabled the Greek mathematicians to make tremendous progress in geometry by supplying the necessary logical foundation for incommensurable ratios." This incommenurability is dealt with in Euclid's Elements, Book X, Proposition 9.

As a result of the distinction between number and magnitude, geometry became the only method that could take into account incommensurable ratios. Because previous numerical foundations were still incompatible with the concept of incommensurability, Greek focus shifted away from those numerical conceptions such as algebra and focused almost exclusively on geometry. In fact, in many cases algebraic conceptions were reformulated into geometrical terms. This may account for why we still conceive of x^2 or x^3 as x squared and x cubed instead of x second power and x third power. Also crucial to Zeno's work with incommensurable magnitudes was the fundamental focus on deductive reasoning that resulted from the foundational shattering of earlier Greek mathematics. The realization that some basic conception within the existing theory was at odds with reality necessitated a complete and thorough investigation of the axioms and assumptions that underlie that theory. Out of this necessity, Eudoxus developed his method of exhaustion, a kind of reductio ad absurdum that "...established the deductive organization on the basis of explicit axioms..." as well as "...reinforced the earlier decision to rely on deductive reasoning for proof." This method of exhaustion is the first step in the creation of calculus.

Theodorus of Cyrene proved the irrationality of the surds of whole numbers up to 17, but stopped there probably because the algebra he used couldn't be applied to the square root of 17.

It wasn't until Eudoxus developed a theory of proportion that took into account irrational as well as rational ratios that a strong mathematical foundation of irrational numbers was created.

India

Geometrical and mathematical problems involving irrational numbers such as square roots were addressed very early during the Vedic period in India and there are references to such calculations in the Samhitas, Brahmanas and more notably in the Sulbha sutras (800 BC or earlier).

It is suggested that the concept of irrationality was implicitly accepted by Indian mathematicians since the 7th century BC, when Manava (c. 750 – 690 BC) believed that the square roots of numbers such as 2 and 61 could not be exactly determined. However, historian Carl Benjamin Boyer writes that "such claims are not well substantiated and unlikely to be true".

It is also suggested that Aryabhata (5th century AD), in calculating a value of pi to 5 significant figures, used the word āsanna (approaching), to mean that not only is this an approximation but that the value is incommensurable (or irrational).

Later, in their treatises, Indian mathematicians wrote on the arithmetic of surds including addition, subtraction, multiplication, rationalization, as well as separation and extraction of square roots..

Mathematicians like Brahmagupta (in 628 AD) and Bhaskara I (in 629 AD) made contributions in this area as did other mathematicians who followed. In the 12th century Bhaskara II evaluated some of these formulas and critiqued them, identifying their limitations.

During the 14th to 16th centuries, Madhava of Sangamagrama and the Kerala school of astronomy and mathematics discovered the infinite series for several irrational numbers such as π and certain irrational values of trigonometric functions. Jyesthadeva provided proofs for these infinite series in the Yuktibhā.

Middle Ages

In the Middle ages, the development of algebra by Muslim mathematicians allowed irrational numbers to be treated as algebraic objects. Middle Eastern mathematicians also merged the concepts of "number" and "magnitude" into a more general idea of real numbers, criticized Euclid's idea of ratios, developed the theory of composite ratios, and extended the concept of number to ratios of continuous magnitude. In his commentary on Book 10 of the Elements, the Persian mathematician Al-Mahani (d. 874/884) examined and classified quadratic irrationals and cubic irrationals. He provided definitions for rational and irrational magnitudes, which he treated as irrational numbers. He dealt with them freely but explains them in geometric terms as follows:

"It will be a rational (magnitude) when we, for instance, say 10, 12, 3%, 6%, etc., because its value is pronounced and expressed quantitatively. What is not rational is irrational and it is impossible to pronounce and represent its value quantitatively. For example: the roots of numbers such as 10, 15, 20 which are not squares, the sides of numbers which are not cubes etc."

In contrast to Euclid's concept of magnitudes as lines, Al-Mahani considered integers and fractions as rational magnitudes, and square roots and cube roots as irrational magnitudes. He also introduced an arithmetical approach to the concept of irrationality, as he attributes the following to irrational magnitudes:

"their sums or differences, or results of their addition to a rational magnitude, or results of subtracting a magnitude of this kind from an irrational one, or of a rational magnitude from it."

The Egyptian mathematician Abū Kāmil Shujā ibn Aslam (c. 850 – 930) was the first to accept irrational numbers as solutions to quadratic equations or as coefficients in an equation, often in the form of square roots, cube roots and fourth roots. In the 10th century, the Iraqi mathematician Al-Hashimi provided general proofs (rather than geometric demonstrations) for irrational numbers, as he considered multiplication, division, and other arithmetical functions. Iranian mathematician, Abū Ja'far al-Khāzin (900–971) provides a definition of rational and irrational magnitudes, stating that if a definite quantity is:

"contained in a certain given magnitude once or many times, then this (given) magnitude corresponds to a rational number. . . . Each time when this (latter) magnitude comprises a half, or a third, or a quarter of the given magnitude (of the unit), or, compared with (the unit), comprises three, five, or three fifths, it is a rational magnitude. And, in general, each magnitude that corresponds to this magnitude (i.e. to the unit), as one number to another, is rational. If, however, a magnitude cannot be represented as a multiple, a part (l/n), or parts (m/n) of a given magnitude, it is irrational, i.e. it cannot be expressed other than by means of roots."

Many of these concepts were eventually accepted by European mathematicians sometime after the Latin translations of the 12th century. Al-Hassār, a Moroccan mathematician from Fez specializing in Islamic inheritance jurisprudence during the 12th century, first mentions the use of a fractional bar, where numerators and denominators are separated by a horizontal bar. In his discussion he writes, "..., for example, if you are told to write three-fifths and a third of a fifth, write thus, $\frac{3 \quad 1}{5 \quad 3}$." This same fractional notation appears soon after in the work of Leonardo Fibonacci in the 13th century.

Modern Period

The 17th century saw imaginary numbers become a powerful tool in the hands of Abraham de Moivre, and especially of Leonhard Euler. The completion of the theory of complex numbers in the 19th century entailed the differentiation of irrationals into algebraic and transcendental numbers, the proof of the existence of transcendental numbers, and the resurgence of the scientific study of the theory of irrationals, largely ignored since Euclid. The year 1872 saw the publication of the theories of Karl Weierstrass (by his pupil Ernst Kossak), Eduard Heine (Crelle's Journal, 74), Georg Cantor (Annalen, 5), and Richard Dedekind. Méray had taken in 1869 the same point of departure as Heine, but the theory is generally referred to the year 1872. Weierstrass's method has been completely set forth by Salvatore Pincherle in 1880, and Dedekind's has received additional prominence through the author's later work (1888) and the endorsement by Paul Tannery (1894). Weierstrass, Cantor, and Heine base their theories on infinite series, while Dedekind founds his on the idea of a cut (Schnitt) in the system of real numbers, separating all rational numbers into two groups having certain characteristic properties. The subject has received later contributions at the hands of Weierstrass, Leopold Kronecker (Crelle, 101), and Charles Méray.

Continued fractions, closely related to irrational numbers (and due to Cataldi, 1613), received attention at the hands of Euler, and at the opening of the 19th century were brought into prominence through the writings of Joseph Louis Lagrange. Dirichlet also added to the general theory, as have numerous contributors to the applications of the subject.

Johann Heinrich Lambert proved (1761) that π cannot be rational, and that e^n is irrational if n is rational (unless n = 0). While Lambert's proof is often called incomplete, modern assessments support it as satisfactory, and in fact for its time it is unusually rigorous. Adrien-Marie Legendre (1794), after introducing the Bessel–Clifford function, provided a proof to show that π^2 is irrational, whence it follows immediately that π is irrational also. The existence of transcendental numbers was first established by Liouville (1844, 1851). Later, Georg Cantor (1873) proved their existence by a different method, that showed that every interval in the reals contains transcendental numbers. Charles

Hermite (1873) first proved e transcendental, and Ferdinand von Lindemann (1882), starting from Hermite's conclusions, showed the same for π. Lindemann's proof was much simplified by Weierstrass (1885), still further by David Hilbert (1893), and was finally made elementary by Adolf Hurwitz and Paul Gordan.

Example Proofs

Square Roots

The square root of 2 was the first number proved irrational, and that article contains a number of proofs. The golden ratio is another famous quadratic irrational and there is a simple proof of its irrationality in its article. The square roots of all natural numbers which are not perfect squares are irrational and a proof may be found in quadratic irrationals.

General Roots

The proof above for the square root of two can be generalized using the fundamental theorem of arithmetic. This asserts that every integer has a unique factorization into primes. Using it we can show that if a rational number is not an integer then no integral power of it can be an integer, as in lowest terms there must be a prime in the denominator that does not divide into the numerator whatever power each is raised to. Therefore, if an integer is not an exact k^{th} power of another integer then its kth root is irrational.

Logarithms

Perhaps the numbers most easy to prove irrational are certain logarithms. Here is a proof by contradiction that $\log_2 3$ is irrational. Notice that $\log_2 3 \approx 1.58 > 0$.

Assume $\log_2 3$ is rational. For some positive integers m and n, we have

$$\log_2 3 = \frac{m}{n}.$$

It follows that

$$2^{m/n} = 3$$

$$(2^{m/n})^n = 3^n$$

$$2^m = 3^n.$$

However, the number 2 raised to any positive integer power must be even (because it is divisible by 2) and the number 3 raised to any positive integer power must be odd (since none of its prime factors will be 2). Clearly, an integer cannot be both odd and even at the same time: we have a contradiction. The only assumption we made was that $\log_2 3$ is rational (and so expressible as a quotient of integers m/n with $n \neq 0$). The contradiction means that this assumption must be false, i.e. $\log_2 3$ is irrational, and can never be expressed as a quotient of integers m/n with $n \neq 0$.

Cases such as $\log_{10} 2$ can be treated similarly.

Transcendental and Algebraic Irrationals

Almost all irrational numbers are transcendental and all real transcendental numbers are irrational (there are also complex transcendental numbers): the article on transcendental numbers lists several examples. e^r and π^r are irrational if $r \neq 0$ is rational; e^π is irrational.

Another way to construct irrational numbers is as irrational algebraic numbers, i.e. as zeros of polynomials with integer coefficients: start with a polynomial equation

$$p(x) = a_n x^n + a_{n-1} x^{n-1} + \cdots + a_1 x + a_0 = 0$$

where the coefficients a_i are integers. Suppose you know that there exists some real number x with p(x) = 0 (for instance if n is odd and a_n is non-zero, then because of the intermediate value theorem). The only possible rational roots of this polynomial equation are of the form r/s where r is a divisor of a_0 and s is a divisor of a_n; there are only finitely so many such candidates you can check by hand. If neither of them is a root of p, then x must be irrational. For example, this technique can be used to show that $x = (2^{1/2} + 1)^{1/3}$ is irrational: we have $(x^3 - 1)^2 = 2$ and hence $x^6 - 2x^3 - 1 = 0$, and this latter polynomial does not have any rational roots (the only candidates to check are ±1).

Because the algebraic numbers form a field, many irrational numbers can be constructed by combining transcendental and algebraic numbers. For example, $3\pi + 2$, $\pi + \sqrt{2}$ and $e\sqrt{3}$ are irrational (and even transcendental).

Decimal Expansions

The decimal expansion of an irrational number never repeats or terminates (essentially, that is repeating zeroes), unlike any rational number does. Similarly for binary, octal or hexadecimal expansions, and in general for expansions in every positional notation with natural bases.

To show this, suppose we divide integers n by m (where m is nonzero). When long division is applied to the division of n by m, only m remainders are possible. If 0 appears as a remainder, the decimal expansion terminates. If 0 never occurs, then the algorithm can run at most m − 1 steps without using any remainder more than once. After that, a remainder must recur, and then the decimal expansion repeats.

Conversely, suppose we are faced with a repeating decimal, we can prove that it is a fraction of two integers. For example, consider:

$$A = 0.7\,162\,162\,162\,\cdots.$$

Here the repetend is 162 and the length of the repetend is 3. First, we multiply by an appropriate power of 10 to move the decimal point to the right so that it is just in front of a repetend. In this example we would multiply by 10 to obtain:

$$10A = 7.162\,162\,162\,\cdots.$$

Now we multiply this equation by 10^r where r is the length of the repetend. This has the effect of moving the decimal point to be in front of the "next" repetend. In our example, multiply by 10^3:

$$10,000A = 7162.162\,162 \cdots.$$

The result of the two multiplications gives two different expressions with exactly the same "decimal portion", that is, the tail end of 10,000A matches the tail end of 10A exactly. Here, both 10,000A and 10A have .162162162 ... at the end.

Therefore, when we subtract the 10A equation from the 10,000A equation, the tail end of 10A cancels out the tail end of 10,000A leaving us with:

$$9990A = 7155.$$

Then

$$A = \frac{7155}{9990},$$

is a ratio of integers and therefore a rational number, as required for the proof.

Irrational Powers

Dov Jarden gave a simple non-constructive proof that there exist two irrational numbers a and b, such that a^b is rational:

Consider $\sqrt{2}^{\sqrt{2}}$; if this is rational, then take a = b = $\sqrt{2}$. Otherwise, take a to be the irrational number $\sqrt{2}^{\sqrt{2}}$ and b = $\sqrt{2}$. Then $a^b = (\sqrt{2}^{\sqrt{2}})^{\sqrt{2}} = \sqrt{2}^{\sqrt{2}\cdot\sqrt{2}} = \sqrt{2}^2 = 2$, which is rational.

Although the above argument does not decide between the two cases, the Gelfond–Schneider theorem shows that $\sqrt{2}^{\sqrt{2}}$ is transcendental, hence irrational. This theorem states that if a and b are both algebraic numbers, and a is not equal to 0 or 1, and b is not a rational number, then any value of a^b is a transcendental number (there can be more than one value if complex number exponentiation is used).

An example that provides a simple constructive proof is

$$\left(\sqrt{2}\right)^{\log_{\sqrt{2}} 3} = 3.$$

The base of the left side is irrational and the right side is rational, so one must prove that the exponent on the left side, $\log_{\sqrt{2}} 3$, is irrational. This is so because, by the formula relating logarithms with different bases,

$$\log_{\sqrt{2}} 3 = \frac{\log_2 3}{\log_2 \sqrt{2}} = \frac{\log_2 3}{1/2} = 2\log_2 3$$

which we can assume, for the sake of establishing a contradiction, equals a ratio m/n of positive integers. Then $\log_2 3 = m/2n$ hence $2^{\log_2 3} = 2^{m/2n}$ hence $3 = 2^{m/2n}$ hence $3^{2n} = 2^m$, which is a contradictory pair of prime factorizations and hence violates the fundamental theorem of arithmetic (unique prime factorization).

A stronger result is the following: Every rational number in the interval $((1/e)^{1/e}, \infty)$ can be written either as a^a for some irrational number a or as n^n for some natural number n. Similarly, every positive rational number can be written either as a^{a^a} for some irrational number a or as n^{n^n} for some natural number n.

Open Questions

It is not known whether $\pi + e$ or $\pi - e$ is irrational or not. In fact, there is no pair of non-zero integers m and n for which it is known whether $m\pi + ne$ is irrational or not. Moreover, it is not known whether the set $\{\pi, e\}$ is algebraically independent over **Q**.

It is not known whether πe, π/e, 2^e, e^e, e^{ee}, π^e, $\pi^{\sqrt{2}}$, ln π, Catalan's constant, or the Euler–Mascheroni gamma constant γ are irrational. It is not known if the tetrations ${}^n\pi$ or ${}^ne$ are rational for any positive integer n.

Set of all Irrationals

Since the reals form an uncountable set, of which the rationals are a countable subset, the complementary set of irrationals is uncountable.

Under the usual (Euclidean) distance function $d(x, y) = |x - y|$, the real numbers are a metric space and hence also a topological space. Restricting the Euclidean distance function gives the irrationals the structure of a metric space. Since the subspace of irrationals is not closed, the induced metric is not complete. However, being a G-delta set—i.e., a countable intersection of open subsets—in a complete metric space, the space of irrationals is completely metrizable: that is, there is a metric on the irrationals inducing the same topology as the restriction of the Euclidean metric, but with respect to which the irrationals are complete. One can see this without knowing the aforementioned fact about G-delta sets: the continued fraction expansion of an irrational number defines a homeomorphism from the space of irrationals to the space of all sequences of positive integers, which is easily seen to be completely metrizable.

Furthermore, the set of all irrationals is a disconnected metrizable space. In fact, the irrationals have a basis of clopen sets so the space is zero-dimensional.

ALGEBRAIC NUMBER

An algebraic number is any complex number that is a root of a non-zero polynomial in one variable with rational coefficients (or equivalently – by clearing denominators – with integer coefficients). All integers and rational numbers are algebraic, as are all roots of integers. The same is not true for all real and complex numbers because of transcendental numbers such as π and e. Almost all real and complex numbers are transcendental.

Examples

- The rational numbers, expressed as the quotient of two integers a and b, b not equal to zero, satisfy the above definition because $x = a/b$ is the root of $bx - a$.
- The quadratic surds (irrational roots of a quadratic polynomial $ax^2 + bx + c$ with integer coefficients a, b, and c) are algebraic numbers. If the quadratic polynomial is monic ($a = 1$) then the roots are quadratic integers.
- The constructible numbers are those numbers that can be constructed from a given unit length using straightedge and compass. These include all quadratic surds, all rational numbers, and all numbers that can be formed from these using the basic arithmetic operations and the extraction of square roots. (Note that by designating cardinal directions for 1, −1, i, and −i, complex numbers such as $3 + \sqrt{2}i$ are considered constructible.)
- Any expression formed from algebraic numbers using any combination of the basic arithmetic operations and extraction of nth roots gives another algebraic number.
- Polynomial roots that cannot be expressed in terms of the basic arithmetic operations and extraction of nth roots (such as the roots of $x^5 - x + 1$). This happens with many, but not all, polynomials of degree 5 or higher.
- Gaussian integers: those complex numbers $a + bi$ where both a and b are integers are also quadratic integers.
- Trigonometric functions of rational multiples of π (except when undefined): that is, the trigonometric numbers. For example, each of $\cos \pi/7$, $\cos 3\pi/7$, $\cos 5\pi/7$ satisfies $8x^3 - 4x^2 - 4x + 1 = 0$. This polynomial is irreducible over the rationals, and so these three cosines are conjugate algebraic numbers. Likewise, $\tan 3\pi/16$, $\tan 7\pi/16$, $\tan 11\pi/16$, $\tan 15\pi/16$ all satisfy the irreducible polynomial $x^4 - 4x^3 - 6x^2 + 4x + 1$, and so are conjugate algebraic integers.
- Some irrational numbers are algebraic and some are not:
- The numbers $\sqrt{2}$ and $\sqrt[3]{3}/2$ are algebraic since they are roots of polynomials $x^2 - 2$ and $8x^3 - 3$, respectively.
- The golden ratio φ is algebraic since it is a root of the polynomial $x^2 - x - 1$.
- The numbers π and e are not algebraic numbers; hence they are transcendental.

Properties

- The set of algebraic numbers is countable (enumerable).
- Hence, the set of algebraic numbers has Lebesgue measure zero (as a subset of the complex numbers), i.e. "almost all" complex numbers are not algebraic.
- Given an algebraic number, there is a unique monic polynomial (with rational coefficients) of least degree that has the number as a root. This polynomial is called its minimal polynomial. If its minimal polynomial has degree n, then the algebraic number is said to be of degree n. An algebraic number of degree 1 is

a rational number. A real algebraic number of degree 2 is a quadratic irrational.

- All algebraic numbers are computable and therefore definable and arithmetical.
- The set of real algebraic numbers is linearly ordered, countable, densely ordered, and without first or last element, so is order-isomorphic to the set of rational numbers.
- For real numbers a and b, the complex number a + bi is algebraic if and only if both a and b are algebraic.

Algebraic numbers on the complex plane colored by degree (red=1, green=2, blue=3, yellow=4)

The Field of Algebraic Numbers

The sum, difference, product and quotient (if the denominator is nonzero) of two algebraic numbers is again algebraic (this fact can be demonstrated using the resultant), and the algebraic numbers therefore form a field **Q** (sometimes denoted by **A**, though this usually denotes the adele ring).

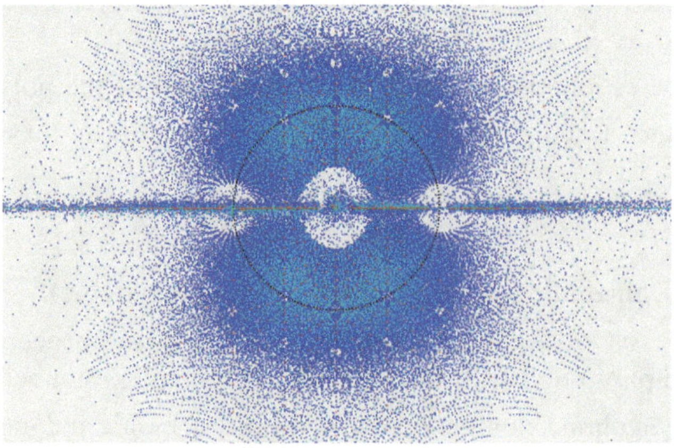

Algebraic numbers colored by degree (blue=4, cyan=3, red=2, green=1). The unit circle is black.

Every root of a polynomial equation whose coefficients are algebraic numbers is again algebraic. This can be rephrased by saying that the field of algebraic numbers is algebraically closed. In fact, it is the smallest algebraically closed field containing the rationals, and is therefore called the algebraic closure of the rationals.

The set of real algebraic numbers itself forms a field.

Related Fields

Numbers Defined by Radicals

All numbers that can be obtained from the integers using a finite number of integer additions, subtractions, multiplications, divisions, and taking nth roots where n is a positive integer (i.e., radical expressions) are algebraic. The converse, however, is not true: there are algebraic numbers that cannot be obtained in this manner. All of these numbers are roots of polynomials of degree ≥ 5. This is a result of Galois theory. An example of such a number is the unique real root of the polynomial $x^5 - x - 1$ (which is approximately 1.167304).

Closed-Form Number

Algebraic numbers are all numbers that can be defined explicitly or implicitly in terms of polynomials, starting from the rational numbers. One may generalize this to "closed-form numbers", which may be defined in various ways. Most broadly, all numbers that can be defined explicitly or implicitly in terms of polynomials, exponentials, and logarithms are called "elementary numbers", and these include the algebraic numbers, plus some transcendental numbers. Most narrowly, one may consider numbers explicitly defined in terms of polynomials, exponentials, and logarithms – this does not include all algebraic numbers, but does include some simple transcendental numbers such as e or log(2).

Algebraic Integers

An algebraic integer is an algebraic number that is a root of a polynomial with integer coefficients with leading coefficient 1 (a monic polynomial). Examples of algebraic integers are $5 + 13\sqrt{2}$, $2 - 6i$, and $1/2(1 + i\sqrt{3})$. Note, therefore, that the algebraic integers constitute a proper superset of the integers, as the latter are the roots of monic polynomials $x - k$ for all $k \in \mathbf{Z}$. In this sense, algebraic integers are to algebraic numbers what integers are to rational numbers.

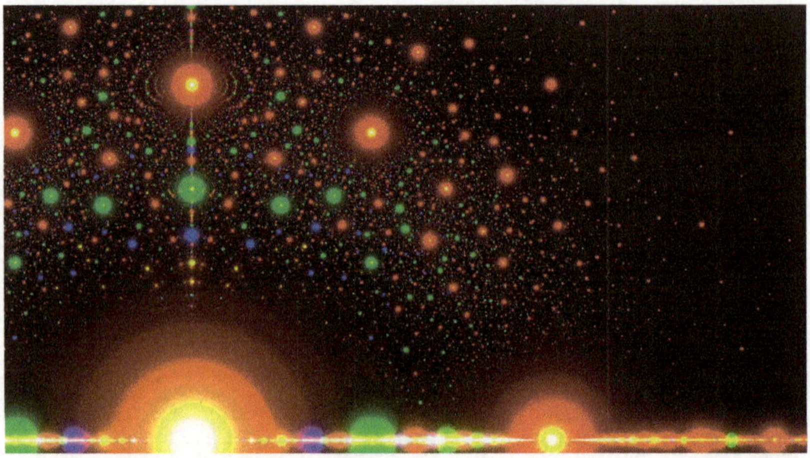

Algebraic numbers colored by leading coefficient (red signifies 1 for an algebraic integer)

The sum, difference and product of algebraic integers are again algebraic integers, which means that the algebraic integers form a ring. The name algebraic integer comes from the fact that the only rational numbers that are algebraic integers are the integers, and because the algebraic integers in any number field are in many ways analogous to the integers. If K is a number field, its ring of integers is the subring of algebraic integers in K, and is frequently denoted as O_K. These are the prototypical examples of Dedekind domains.

TRANSCENDENTAL NUMBER

In mathematics, a transcendental number is a real or complex number that is not algebraic—that is, it is not a root of a non-zero polynomial equation with integer (or, equivalently, rational) coefficients. The best-known transcendental numbers are π and e. Though only a few classes of transcendental numbers are known (in part because it can be extremely difficult to show that a given number is transcendental), transcendental numbers are not rare. Indeed, almost all real and complex numbers are transcendental, since the algebraic numbers are countable while the sets of real and complex numbers are both uncountable. All real transcendental numbers are irrational, since all rational numbers are algebraic. The converse is not true: not all irrational numbers are transcendental; e.g., the square root of 2 is irrational but not a transcendental number, since it is a solution of the polynomial equation x2 − 2 = 0. Another irrational number that is not transcendental is the golden ratio, φ or ϕ..

History

The name "transcendental" comes from the root trans meaning across and length of numbers and from Leibniz's 1682 paper in which he proved that sin(x) is not an algebraic function of x. Euler was probably the first person to define transcendental numbers in the modern sense. Johann Heinrich Lambert conjectured that e and π were both transcendental numbers in his 1768 paper proving the number π is irrational, and proposed a tentative sketch of a proof of π's transcendence.

Joseph Liouville first proved the existence of transcendental numbers in 1844, and in 1851 gave the first decimal examples such as the Liouville constant

$$\sum_{k=1}^{\infty} 10^{-k!} = 0.110001000000000000000001000\ldots$$

in which the nth digit after the decimal point is 1 if n is equal to k! (k factorial) for some k and 0 otherwise. In other words, the nth digit of this number is 1 only if nis one of the numbers 1! = 1, 2! = 2, 3! = 6, 4! = 24, etc. Liouville showed that this number is what we now call a Liouville number; this essentially means that it can be more closely approximated by rational numbers than can any irrational algebraic number. Liouville showed that all Liouville numbers are transcendental.

The first number to be proven transcendental without having been specifically constructed for the purpose was e, by Charles Hermite in 1873.

In 1874, Georg Cantor proved that the algebraic numbers are countable and the real numbers are uncountable. He also gave a new method for constructing transcendental numbers. In 1878, Cantor published a construction that proves there are as many transcendental numbers as there are real numbers.Cantor's work established the ubiquity of transcendental numbers.

In 1882, Ferdinand von Lindemann published a proof that the number π is transcendental. He first showed that e^a is transcendental when a is algebraic and not zero. Then, since $e^{i\pi} = -1$ is algebraic, $i\pi$ and therefore π must be transcendental. This approach was generalized by Karl Weierstrass to the Lindemann–Weierstrass theorem. The transcendence of π allowed the proof of the impossibility of several ancient geometric constructions involving compass and straightedge, including the most famous one, squaring the circle.

In 1900, David Hilbert posed an influential question about transcendental numbers, Hilbert's seventh problem: If a is an algebraic number, that is not zero or one, and b is an irrational algebraic number, is a^b necessarily transcendental? The affirmative answer was provided in 1934 by the Gelfond–Schneider theorem. This work was extended by Alan Baker in the 1960s in his work on lower bounds for linear forms in any number of logarithms (of algebraic numbers).

Properties

The set of transcendental numbers is uncountably infinite. Since the polynomials with rational coefficients are countable, and since each such polynomial has a finite number of zeroes, the algebraic numbers must also be countable. However, Cantor's diagonal argument proves that the real numbers (and therefore also the complex numbers) are uncountable. Since the real numbers are the union of algebraic and transcendental numbers, they cannot both be countable. This makes the transcendental numbers uncountable.

No rational number is transcendental and all real transcendental numbers are irrational. The irrational numbers contain all the real transcendental numbers and a subset of the algebraic numbers, including the quadratic irrationals and other forms of algebraic irrationals.

Any non-constant algebraic function of a single variable yields a transcendental value when applied to a transcendental argument. For example, from knowing that π is transcendental, we can immediately deduce that numbers such as 5π, $(\pi - 3)/\sqrt{2}$, $(\sqrt{\pi} - \sqrt{3})^8$ and $(\pi^5 + 7)^{1/7}$ are transcendental as well.

However, an algebraic function of several variables may yield an algebraic number when applied to transcendental numbers if these numbers are not algebraically independent. For example, π and $(1 - \pi)$ are both transcendental, but $\pi + (1 - \pi) = 1$ is obviously not. It is unknown whether $\pi + e$, for example, is transcendental, though at least one of $\pi + e$ and πe must be transcendental. More generally, for any two transcendental numbers a and b, at least one of a + b and ab must be transcendental. To see this, consider the polynomial $(x - a)(x - b) = x^2 - (a + b)x + ab$. If $(a + b)$ and ab were both algebraic, then this would be a polynomial with algebraic coefficients. Because algebraic numbers form an algebraically closed field, this would imply that the roots of the polynomial, a and b, must be algebraic. But this is a contradiction, and thus it must be the case that at least one of the coefficients is transcendental.

The non-computable numbers are a strict subset of the transcendental numbers.

All Liouville numbers are transcendental, but not vice versa. Any Liouville number must have unbounded partial quotients in its continued fraction expansion. Using a counting argument one can show that there exist transcendental numbers which have bounded partial quotients and hence are not Liouville numbers.

Using the explicit continued fraction expansion of e, one can show that e is not a Liouville number (although the partial quotients in its continued fraction expansion are unbounded). Kurt Mahler showed in 1953 that π is also not a Liouville number. It is conjectured that all infinite continued fractions with bounded terms that are not eventually periodic are transcendental (eventually periodic continued fractions correspond to quadratic irrationals).

Numbers Proven to be Transcendental

Numbers proven to be transcendental:

- e^a if a is algebraic and nonzero (by the Lindemann–Weierstrass theorem).
- π (by the Lindemann–Weierstrass theorem).
- e^{π}, Gelfond's constant, as well as $e^{-\pi/2}=i^{\,i}$ (by the Gelfond–Schneider theorem).
- a^b where a is algebraic but not 0 or 1, and b is irrational algebraic (by the Gelfond–Schneider theorem), in particular:

$2^{\sqrt{2}}$ the Gelfond–Schneider constant (or Hilbert number).

- The Continued Fraction Constant, Carl Ludwig Siegel (1929)

$$1+\cfrac{1}{2+\cfrac{1}{3+\cfrac{1}{4+\cfrac{1}{5+\cfrac{1}{6+\ddots}}}}}$$

- sin(a), cos(a) and tan(a), and their multiplicative inverses csc(a), sec(a) and cot(a), for any nonzero algebraic number a (by the Lindemann–Weierstrass theorem).
- The fixed point of the cosine function - the unique real solution to the equation $\cos(x) = x$ (by the Lindemann–Weierstrass theorem).
- ln(a) if a is algebraic and not equal to 0 or 1, for any branch of the logarithm function (by the Lindemann–Weierstrass theorem).
- W(a) if a is algebraic and nonzero, for any branch of the Lambert W Function (by the Lindemann–Weierstrass theorem).
- Γ(1/3), Γ(1/4), and Γ(1/6).
- 0.64341054629..., Cahen's constant.
- 0.12345678910111213141516..., the Champernowne constant.
- Ω, Chaitin's constant (since it is a non-computable number).
- The Fredholm number

$$\sum_{n=0}^{\infty} 2^{-2^n}$$

more generally, any number of the form

$$\sum_{n=0}^{\infty} \beta^{2^n}$$

with $0 < |\beta| < 1$ and β algebraic.

- The aforementioned Liouville constant

$$\sum_{n=1}^{\infty} 10^{-n!};$$

more generally any number of the form

$$\sum_{n=1}^{\infty} \beta^{n!}$$

with $0 < |\beta| < 1$ and β algebraic.

- The Prouhet–Thue–Morse constant.
- Any number for which the digits with respect to some fixed base form a Sturmian word.
- For $\beta > 1$

$$\sum_{k=0}^{\infty} 10^{-\lfloor \beta^k \rfloor};$$

where $\beta \mapsto \lfloor \beta \rfloor$ is the floor function.

- 3.300330000000000330033... and its reciprocal 0.30300000303..., two numbers with only two different decimal digits whose nonzero digit positions are given by the Moser–de Bruijn sequence and its double.

Possibly Transcendental Numbers

Numbers which have yet to be proven to be either transcendental or algebraic:

- Most sums, products, powers, etc. of the number π and the number e, e.g. $\pi + e$, $\pi - e$, πe, π/e, π^π, e^e, π^e, $\pi^{\sqrt{2}}$, $e^{\pi 2}$ are not known to be rational, algebraic, irrational or transcendental. Notable exceptions are $\pi + e^\pi$, πe^π and $e^{\pi\sqrt{n}}$ (for any positive integer n) which have been proven to be transcendental.
- The Euler–Mascheroni constant γ (which has not even been proven to be irrational).
- Catalan's constant, also not known to be irrational.
- Apéry's constant, ζ(3) (which Apéry proved is irrational)
- The Riemann zeta function at other odd integers, ζ(5), ζ(7), ... (not known to be irrational.)
- The Feigenbaum constants, δ and α.
- Mills' constant.

Conjectures:

- Schanuel's conjecture,
- Four exponentials conjecture.

Sketch of a Proof that e is Transcendental

The first proof that the base of the natural logarithms, e, is transcendental dates from 1873. We will now follow the strategy of David Hilbert (1862–1943) who gave a simplification of the original proof of Charles Hermite. The idea is the following:

Assume, for purpose of finding a contradiction, that e is algebraic. Then there exists a finite set of integer coefficients c_0, c_1, ..., c_n satisfying the equation:

$$c_0 + c_1 e + c_2 e^2 + \cdots + c_n e^n = 0, \qquad c_0, c_n \neq 0.$$

Now for a positive integer k, we define the following polynomial:

$$f_k(x) = x^k[(x-1)\cdots(x-n)]^{k+1},$$

and multiply both sides of the above equation by

$$\int_0^\infty f_k e^{-x}\, dx,$$

to arrive at the equation:

$$c_0\left(\int_0^\infty f_k e^{-x}\, dx\right) + c_1 e\left(\int_0^\infty f_k e^{-x}\, dx\right) + \cdots + c_n e^n\left(\int_0^\infty f_k e^{-x}\, dx\right) = 0.$$

This equation can be written in the form

$$P + Q = 0$$

where

$$P = c_0\left(\int_0^\infty f_k e^{-x}\, dx\right) + c_1 e\left(\int_1^\infty f_k e^{-x}\, dx\right) + c_2 e^2\left(\int_2^\infty f_k e^{-x}\, dx\right) + \cdots + c_n e^n\left(\int_n^\infty f_k e^{-x}\, dx\right)$$

$$Q = c_1 e\left(\int_0^1 f_k e^{-x}\, dx\right) + c_2 e^2\left(\int_0^2 f_k e^{-x}\, dx\right) + \cdots + c_n e^n\left(\int_0^n f_k e^{-x}\, dx\right)$$

Lemma 1. For an appropriate choice of k, $\frac{P}{k!}$ is a non-zero integer.

Proof. Each term in P is an integer times a sum of factorials, which results from the relation

$$\int_0^\infty x^j e^{-x}\, dx = j!$$

which is valid for any positive integer j (consider the Gamma function).

It is non-zero because for every a satisfying $0 < a \leq n$, the integrand in

$$c_a e^a \int_a^\infty f_k e^{-x}\, dx$$

is e^{-x} times a sum of terms whose lowest power of x is k+1 after substituting x for x - a in the integral. Then this becomes a sum of integrals of the form

$$\int_0^{\infty} x^j e^{-x}\,dx$$

with k+1 ≤ j, and it is therefore an integer divisible by (k+1)!. After dividing by k!, we get zero modulo (k+1). However, we can write:

$$\int_0^{\infty} f_k e^{-x}\,dx = \int_0^{\infty} \left([(-1)^n (n!)]^{k+1} e^{-x} x^k + \cdots\right) dx$$

and thus

$$\frac{1}{k!} c_0 \int_0^{\infty} f_k e^{-x}\,dx = c_0 [(-1)^n (n!)]^{k+1} \qquad \bmod (k+1).$$

By choosing k so that k+1 is prime and larger than n and $|c_0|$, we get that $\frac{P}{k!}$ is non-zero modulo (k+1) and is thus non-zero.

Lemma 2. $\left|\frac{Q}{k!}\right| < 1$ for sufficiently large k.

Proof. Note that

$$f_k e^{-x} = x^k [(x-1)(x-2)\cdots(x-n)]^{k+1} e^{-x} = \left([x(x-1)\cdots(x-n)]^k\right)\left((x-1)\cdots(x-n)e^{-x}\right)$$

Using upper bounds G and H for $|x(x-1)\cdots(x-n)|$ and $|(x-1)\cdots(x-n)e^{-x}|$ on the interval [0,n] we can infer that

$$|Q| < G^k H(|c_1|e + 2|c_2|e^2 + \cdots + n|c_n|e^n)$$

and since

$$\lim_{k\to\infty} \frac{G^k}{k!} = 0$$

it follows that

$$\lim_{k\to\infty} \frac{Q}{k!} = 0$$

which is sufficient to finish the proof of this lemma.

Noting that one can choose k so that both Lemmas hold we get the contradiction we needed to prove the transcendence of e.

The Transcendence of π

A similar strategy, different from Lindemann's original approach, can be used to show that the number π is transcendental. Besides the gamma-function and some estimates as in the proof for e, facts about symmetric polynomials play a vital role in the proof.

For detailed information concerning the proofs of the transcendence of π and e see the references and external links.

Mahler's Classification

Kurt Mahler in 1932 partitioned the transcendental numbers into 3 classes, called **S**, **T**, and **U**. Definition of these classes draws on an extension of the idea of a Liouville number (cited above).

Measure of Irrationality of a Real Number

One way to define a Liouville number is to consider how small a given real number **x** makes linear polynomials $|qx - p|$ without making them exactly 0. Here p, q are integers with $|p|$, $|q|$ bounded by a positive integer H.

Let m(x, 1, H) be the minimum non-zero absolute value these polynomials take and take:

$$\omega(x,1,H) = -\frac{\log m(x,1,H)}{\log H}$$

$$\omega(x,1) = \limsup_{H\to\infty} \omega(x,1,H).$$

ω(x, 1) is often called the measure of irrationality of a real number x. For rational numbers, ω(x, 1) = 0 and is at least 1 for irrational real numbers. A Liouville number is defined to have infinite measure of irrationality. Roth's theorem says that irrational real algebraic numbers have measure of irrationality 1.

Measure of Transcendence of a Complex Number

Next consider the values of polynomials at a complex number x, when these polynomials have integer coefficients, degree at most n, and height at most H, with n, H being positive integers.

Let m(x,n,H) be the minimum non-zero absolute value such polynomials take at x and take:

$$\omega(x,n,H) = -\frac{\log m(x,n,H)}{n\log H}$$

$$\omega(x,n) = \limsup_{H\to\infty} \omega(x,n,H).$$

Suppose this is infinite for some minimum positive integer n. A complex number x in this case is called a **U number** of degree n.

Now we can define

$$\omega(x) = \limsup_{n\to\infty} \omega(x,n).$$

ω(x) is often called the measure of transcendence of x. If the ω(x,n) are bounded, then ω(x) is finite, and x is called an S number. If the ω(x,n) are finite but unbounded, x is called a T number. x is algebraic if and only if ω(x) = 0.

Clearly the Liouville numbers are a subset of the U numbers. William LeVeque in 1953 constructed U numbers of any desired degree. The Liouville numbers and hence the U numbers are uncountable sets. They are sets of measure 0.

T numbers also comprise a set of measure 0. It took about 35 years to show their existence. Wolfgang M. Schmidt in 1968 showed that examples exist. However, almost all complex numbers are S numbers. Mahler proved that the exponential function sends all non-zero algebraic numbers to S numbers: this shows that e is an S number and gives a proof of the transcendence of π. The most that is known about π is that it is not a U number. Many other transcendental numbers remain unclassified.

Two numbers x, y are called algebraically dependent if there is a non-zero polynomial P in 2 indeterminates with integer coefficients such that P(x, y) = 0. There is a powerful theorem that 2 complex numbers that are algebraically dependent belong to the same Mahler class. This allows construction of new transcendental numbers, such as the sum of a Liouville number with e or π.

It is often speculated that S stood for the name of Mahler's teacher Carl Ludwig Siegel and that T and U are just the next two letters.

Koksma's Equivalent Classification

Jurjen Koksma in 1939 proposed another classification based on approximation by algebraic numbers.

Consider the approximation of a complex number x by algebraic numbers of degree $\leq$ n and height $\leq$ H. Let α be an algebraic number of this finite set such that $|x - \alpha|$ has the minimum positive value. Define $\omega^*(x,H,n)$ and $\omega^*(x,n)$ by:

$$|x-\alpha| = H^{-n\omega^*(x,H,n)-1}.$$

$$\omega^*(x,n) = \limsup_{H\to\infty} \omega^*(x,n,H).$$

If for a smallest positive integer n, $\omega^*(x,n)$ is infinite, x is called a U*-number of degree n.

If the $\omega^*(x,n)$ are bounded and do not converge to 0, x is called an S*-number,

A number x is called an **A*-number** if the $\omega^*(x,n)$ converge to 0.

If the $\omega^*(x,n)$ are all finite but unbounded, x is called a **T*-number**,

Koksma's and Mahler's classifications are equivalent in that they divide the transcendental numbers into the same classes. The A*-numbers are the algebraic numbers.

LeVeque's Construction

Let

$$\lambda = \frac{1}{3} + \sum_{k=1}^{\infty} 10^{-k!}$$

It can be shown that the nth root of λ (a Liouville number) is a U-number of degree n.

This construction can be improved to create an uncountable family of U-numbers of degree n. Let Z be the set consisting of every other power of 10 in the series above for λ. The set of all subsets of Z is uncountable. Deleting any of the subsets of Z from the series for λ creates uncountably many distinct Liouville numbers, whose nth roots are U-numbers of degree n.

Type

The supremum of the sequence $\{\omega(x, n)\}$ is called the **type**. Almost all real numbers are S numbers of type 1, which is minimal for real S numbers. Almost all complex numbers are S numbers of type 1/2, which is also minimal. The claims of almost all numbers were conjectured by Mahler and in 1965 proved by Vladimir Sprindzhuk.

p-adic NUMBER

In mathematics the ***p***-adic number system for any prime number p extends the ordinary arithmetic of the rational numbers in a different way from the extension of the rational number system to the real and complex number systems. The extension is achieved by an alternative interpretation of the concept of "closeness" or absolute value. In particular, p-adic numbers have the interesting property that they are said to be close when their difference is divisible by a high power of p: the higher the power, the closer they are. This property enables p-adic numbers to encode congruence information in a way that turns out to have powerful applications in number theory—including, for example, in the famous proof of Fermat's Last Theorem by Andrew Wiles.

p-adic numbers were first described by Kurt Hensel in 1897, though, with hindsight, some of Ernst Kummer's earlier work can be interpreted as implicitly using p-adic numbers. The p-adic numbers were motivated primarily by an attempt to bring the ideas and techniques of power series methods into number theory. Their influence now extends far beyond this. For example, the field of p-adic analysis essentially provides an alternative form of calculus.

The 3-adic integers, with selected corresponding characters on their Pontryagin dual group.

More formally, for a given prime p, the field $\mathbf{Q}_p$ of p-adic numbers is a completion of the rational numbers. The field $\mathbf{Q}_p$ is also given a topology derived from a metric, which is itself derived from the p-adic order, an alternative valuation on the rational numbers. This metric space is complete in the sense that every Cauchy sequence converges to a point in $\mathbf{Q}_p$. This is what allows the development of calculus on $\mathbf{Q}_p$, and it is the interaction of this analytic and algebraic structure that gives the p-adic number systems their power and utility.

The p in "p-adic" is a variable and may be replaced with a prime (yielding, for instance, "the 2-adic numbers") or another placeholder variable (for expressions such as "the ℓ-adic numbers"). The "adic" of "p-adic" comes from the ending found in words such as dyadic or triadic.

Introduction

This section is an informal introduction to p-adic numbers, using examples from the ring of 10-adic (decadic) numbers. Although for p-adic numbers p should be a prime, base 10 was chosen to highlight the analogy with decimals. The decadic numbers are generally not used in mathematics: since 10 is not prime, the decadics are not a field. More formal constructions and properties are given below.

In the standard decimal representation, almost all real numbers do not have a terminating decimal representation. For example, 1/3 is represented as a non-terminating decimal as follows

$$\frac{1}{3} = 0.333333\ldots.$$

Informally, non-terminating decimals are easily understood, because it is clear that a real number can be approximated to any required degree of precision by a terminating decimal. If two decimal expansions differ only after the 10th decimal place, they are quite close to one another; and if they differ only after the 20th decimal place, they are even closer.

10-adic numbers use a similar non-terminating expansion, but with a different concept of "closeness". Whereas two decimal expansions are close to one another if their difference is a large negative power of 10, two 10-adic expansions are close if their difference is a large positive power of 10. Thus 4739 and 5739, which differ by 10^3, are close in the 10-adic world, and 72694473 and 82694473 are even closer, differing by 10^7.

More precisely, a positive rational number r can be expressed as $10^e \cdot p/q$, where p and q are positive integers and q is relatively prime to p and to 10. For each $r \neq 0$ there exists the maximal e such that this representation is possible. Let the 10-adic absolute value of r be

$$|r|_{10} = \frac{1}{10^e}$$

$$|0|_{10} = 0.$$

Closeness in any number system is defined by a metric. Using the 10-adic metric the distance between numbers x and y is given by $|x-y|_{10}$. An interesting consequence of the 10-adic metric (or of a p-adic metric) is that there is no longer a need for the negative sign. As an example, by examining the following sequence we can see how unsigned 10-adics can get progressively closer and closer to the number -1:

$$9 = -1 + 10 \quad \text{so } |9 - (-1)|_{10} = \frac{1}{10}.$$

$$99 = -1 + 10^2 \quad \text{so } |99 - (-1)|_{10} = \frac{1}{100}.$$

$$999 = -1 + 10^3 \quad \text{so } |999 - (-1)|_{10} = \frac{1}{1000}.$$

$$9999 = -1 + 10^4 \quad \text{so } |9999 - (-1)|_{10} = \frac{1}{10000}.$$

and taking this sequence to its limit, we can say that the 10-adic expansion of -1 is

$$\ldots 9999 = -1.$$

In this notation, 10-adic expansions can be extended indefinitely to the left, in contrast to decimal expansions, which can be extended indefinitely to the right. Note that this is not the only way to write p-adic numbers – for alternatives see the Notation section below.

More formally, a 10-adic number can be defined as

$$\sum_{i=n}^{\infty} a_i 10^i$$

where each of the a_i is a digit taken from the set $\{0, 1, \ldots, 9\}$ and the initial index n may be positive, negative or 0, but must be finite. From this definition, it is clear that positive integers and positive rational numbers with terminating decimal expansions will have terminating 10-adic expansions that are identical to their decimal expansions. Other numbers may have non-terminating 10-adic expansions.

It is possible to define addition, subtraction, and multiplication on 10-adic numbers in a consistent way, so that the 10-adic numbers form a commutative ring.

We can create 10-adic expansions for negative numbers as follows

$$-100 = -1 \times 100 = \ldots 9999 \times 100 = \ldots 9900$$
$$\Rightarrow -35 = -100 + 65 = \ldots 9900 + 65 = \ldots 9965$$
$$\Rightarrow -\left(3 + \frac{1}{2}\right) = \frac{-35}{10} = \frac{\ldots 9965}{10} = \ldots 9996.5$$

and fractions which have non-terminating decimal expansions also have non-terminating 10-adic expansions. For example

$$\frac{10^6 - 1}{7} = 142857; \frac{10^{12} - 1}{7} = 142857142857; \frac{10^{18} - 1}{7} = 142857142857142857$$
$$\Rightarrow -\frac{1}{7} = \ldots 142857142857142857$$
$$\Rightarrow -\frac{6}{7} = \ldots 142857142857142857 \times 6 = \ldots 857142857142857142$$
$$\Rightarrow \frac{1}{7} = -\frac{6}{7} + 1 = \ldots 857142857142857143.$$

Generalizing the last example, we can find a 10-adic expansion with no digits to the right of the decimal point for any rational number p⁄q such that q is co-prime to 10; Euler's theorem guarantees that if q is co-prime to 10, then there is an n such that $10^n - 1$ is a multiple of q. The other rational numbers can be expressed as 10-adic numbers with some digits after the decimal point.

As noted above, 10-adic numbers have a major drawback. It is possible to find pairs of non-zero 10-adic numbers (having an infinite number of digits, and thus not rational) whose product is 0. This means that 10-adic numbers do not always have multiplicative inverses i.e. valid reciprocals, which in turn implies that though 10-adic numbers form a ring they do not form a field, a deficiency that makes them much less useful as an analytical tool. Another way of saying this is that the ring of 10-adic numbers is not an integral domain because they contain zero divisors. The reason for this property turns out to be that 10 is a composite number which is not a power of a prime. This problem is simply avoided by using a prime number p as the base of the number system instead of 10 and indeed for this reason p in p-adic is usually taken to be prime.

p-adic Expansions

When dealing with natural numbers, if we take p to be a fixed prime number, then any positive integer can be written as a base p expansion in the form

$$\sum_{i=0}^{n} a_i p^i$$

where the a_i are integers in $\{0, \ldots, p-1\}$. For example, the binary expansion of 35 is $1 \cdot 2^5 + 0 \cdot 2^4 + 0 \cdot 2^3 + 0 \cdot 2^2 + 1 \cdot 2^1 + 1 \cdot 2^0$, often written in the shorthand notation 100011_2.

The familiar approach to extending this description to the larger domain of the rationals (and, ultimately, to the reals) is to use sums of the form:

$$\pm \sum_{i=-\infty}^{n} a_i p^i.$$

A definite meaning is given to these sums based on Cauchy sequences, using the absolute value as metric. Thus, for example, 1/3 can be expressed in base 5 as the limit of the sequence $0.1313131313\ldots_5$. In this formulation, the integers are precisely those numbers for which $a_i = 0$ for all $i < 0$.

With p-adic numbers, on the other hand, we choose to extend the base p expansions in a different way. Unlike traditional integers, where the magnitude is determined by how far they are from zero, the "size" of p-adic numbers is determined by the p-adic absolute value, where high positive powers of p are relatively small compared to high negative powers of p. Consider infinite sums of the form:

$$\sum_{i=k}^{\infty} a_i p^i$$

where k is some (not necessarily positive) integer, and each coefficient ai can be called a ***p***-adic digit. With this approach we obtain the p-adic expansions of the p-adic numbers. Those p-adic numbers for which $a_i = 0$ for all $i < 0$ are also called the p-adic integers.

As opposed to real number expansions which extend to the right as sums of ever smaller, increasingly negative powers of the base p, p-adic numbers may expand to the left forever, a property that can often be true for the p-adic integers. For example, consider the p-adic expansion of 1/3 in base 5. It can be shown to be $\ldots 1313132_5$, i.e., the limit of the sequence 2_5, 32_5, 132_5, 3132_5, 13132_5, 313132_5, 1313132_5, … :

$$\frac{5^2-1}{3} = \frac{44_5}{3} = 13_5; \quad \frac{5^4-1}{3} = \frac{4444_5}{3} = 1313_5$$

$$\Rightarrow -\frac{1}{3} = \ldots 1313_5$$

$$\Rightarrow -\frac{2}{3} = \ldots 1313_5 \times 2 = \ldots 3131_5$$

$$\Rightarrow \frac{1}{3} = -\frac{2}{3} + 1 = \ldots 3132_5.$$

Multiplying this infinite sum by 3 in base 5 gives $\ldots 0000001_5$. As there are no negative powers of 5 in this expansion of 1/3 (i.e. no numbers to the right of the decimal point), we see that 1/3 satisfies the definition of being a p-adic integer in base 5.

More formally, the p-adic expansions can be used to define the field $\mathbf{Q}_p$ of ***p*-adic numbers** while the p-adic integers form a subring of $\mathbf{Q}_p$, denoted $\mathbf{Z}_p$. (Not to be confused with the ring of integers modulo p which is also sometimes written $\mathbf{Z}_p$. To avoid ambiguity, $\mathbf{Z}/p\mathbf{Z}$ or $\mathbf{Z}/(p)$ are often used to represent the integers modulo p.)

While it is possible to use the approach above to define p-adic numbers and explore their properties, just as in the case of real numbers other approaches are generally preferred. Hence we want to define a notion of infinite sum which makes these expressions meaningful, and this is most easily accomplished by the introduction of the p-adic metric. Two different but equivalent solutions to this problem are presented in the Constructions section below.

Notation

There are several different conventions for writing p-adic expansions. So far this article has used a notation for p-adic expansions in which powers of p increase from right to left. With this right-to-left notation the 3-adic expansion of $1/5$, for example, is written as

$$\frac{1}{5} = \ldots 121012102_3.$$

When performing arithmetic in this notation, digits are carried to the left. It is also possible to write p-adic expansions so that the powers of p increase from left to right, and digits are carried to the right. With this left-to-right notation the 3-adic expansion of $1/5$ is

$$\frac{1}{5} = 2.01210121\ldots_3 \text{ or } \frac{1}{15} = 20.1210121\ldots_3.$$

p-adic expansions may be written with other sets of digits instead of $\{0, 1, \ldots, p-1\}$. For example, the 3-adic expansion of $1/5$ can be written using balanced ternary digits $\{\underline{1},0,1\}$ as

$$\frac{1}{5} = \ldots \underline{1}11\underline{1}11\underline{1}11\underline{1}_3.$$

In fact any set of p integers which are in distinct residue classes modulo p may be used as p-adic digits. In number theory, Teichmüller representatives are sometimes used as digits.

Constructions

Analytic Approach

$p = 2$		← distance = 1 →							
De-ci-mal	Bi-nary	← d = ½ →				← d = ½ →			
		‹d=¼›		‹d=¼›		‹d=¼›		‹d=¼›	
		‹⅛›	‹⅛›	‹⅛›	‹⅛›	‹⅛›	‹⅛›	‹⅛›	‹⅛›
		...	...	...	...	...	...	...	...
17	10001					J			
16	10000	J							
15	1111								L
14	1110				L				
13	1101						L		
12	1100		L						
11	1011							L	
10	1010			L					
9	1001					L			
8	1000	L							
7	111								L
6	110				L				
5	101						L		
4	100		L						
3	11							L	
2	10			L					
1	1					L			
0	0...000	L							
−1	1...111								J
−2	1...110				J				
−3	1...101						J		
−4	1...100		J						
Dec	Bin	...	...	...	...	...	...	...	...

2-adic (p = 2) arrangement of integers, from left to right. This shows a hierarchical subdivision pattern common for ultrametric spaces. Points within a distance 1/8 are grouped in one colored strip. A pair of strips within a distance 1/4 has the same chroma, four strips within a distance 1/2 have the same hue. The hue is determined by the least significant bit, the saturation – by the next (2^1) bit, and the brightness depends on the value of 2^2 bit. Bits (digit places) which are less significant for the usual metric are more significant for the p-adic distance.

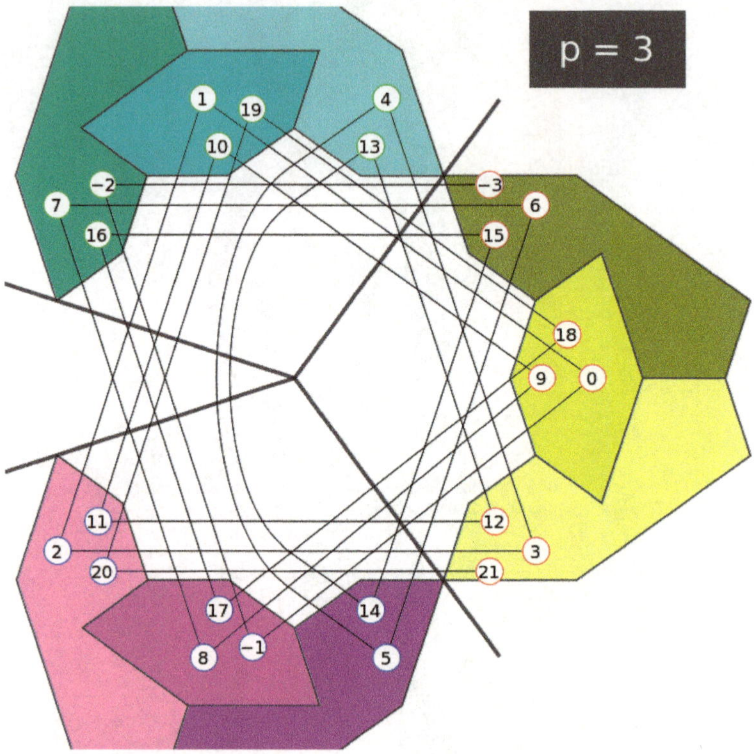

Similar picture for p = 3 (click to enlarge) shows three closed balls of radius 1/3, where each consists of 3 balls of radius 1/9

The real numbers can be defined as equivalence classes of Cauchy sequences of rational numbers; this allows us to, for example, write 1 as 1.000... = 0.999... . The definition of a Cauchy sequence relies on the metric chosen, though, so if we choose a different one, we can construct numbers other than the real numbers. The usual metric which yields the real numbers is called the Euclidean metric.

For a given prime p, we define the p-adic absolute value in **Q** as follows: for any non-zero rational number x, there is a unique integer n allowing us to write $x = p^n(a/b)$, where neither of the integers a and b is divisible by p. Unless the numerator or denominator of x in lowest terms contains p as a factor, n will be 0. Now define $|x|_p = p^{-n}$. We also define $|0|_p = 0$.

For example with $x = 63/550 = 2^{-1}\cdot 3^{2}\cdot 5^{-2}\cdot 7\cdot 11^{-1}$

$$|x|_2 = 2$$

$$|x|_3 = 1/9$$

$$|x|_5 = 25$$

$$|x|_7 = 1/7$$

$$|x|_{11} = 11$$

$$|x|_{\text{any other prime}} = 1.$$

This definition of $|x|_p$ has the effect that high powers of p become "small". By the fundamental theorem of arithmetic, for a given non-zero rational number x there is a unique finite set of distinct primes $p_1, \ldots, p_r$ and a corresponding sequence of non-zero integers $a_1, \ldots, a_r$ such that:

$$|x| = p_1^{a_1} \ldots p_r^{a_r}.$$

It then follows that $|x|_{p_i} = p_i^{-a_i}$ for all $1 \leq i \leq r$, and $|x|_p = 1$ for any other prime $p \notin \{p_1, \ldots, p_r\}$.

The p-adic absolute value defines a metric d_p on $\mathbf{Q}$ by setting

$$d_p(x, y) = |x - y|_p$$

The field $\mathbf{Q}_p$ of p-adic numbers can then be defined as the completion of the metric space $(\mathbf{Q}, d_p)$; its elements are equivalence classes of Cauchy sequences, where two sequences are called equivalent if their difference converges to zero. In this way, we obtain a complete metric space which is also a field and contains $\mathbf{Q}$.

It can be shown that in $\mathbf{Q}_p$, every element x may be written in a unique way as

$$\sum_{i=k}^{\infty} a_i p^i$$

where k is some integer such that $a_k \neq 0$ and each a_i is in $\{0, \ldots, p-1\}$. This series converges to x with respect to the metric d_p.

With this absolute value, the field $\mathbf{Q}_p$ is a local field.

Ostrowski's theorem states that each absolute value on $\mathbf{Q}$ is equivalent either to the Euclidean absolute value, the trivial absolute value, or to one of the p-adic absolute values for some prime p. Each absolute value (or metric) leads to a different completion of $\mathbf{Q}$. (With the trivial absolute value, $\mathbf{Q}$ is already complete.)

Algebraic Approach

In the algebraic approach, we first define the ring of p-adic integers, and then construct the field of fractions of this ring to get the field of p-adic numbers.

We start with the inverse limit of the rings $\mathbf{Z}/p^n\mathbf{Z}$: a p-adic integer is then a sequence $(a_n)_{n\geq 1}$ such that a_n is in $\mathbf{Z}/p^n\mathbf{Z}$, and if $n \leq m$, then $a_n \equiv a_m \pmod{p^n}$.

Every natural number m defines such a sequence (a_n) by $a_n = m \bmod p^n$ and can therefore be regarded as a p-adic integer. For example, in this case 35 as a 2-adic integer would be written as the sequence (1, 3, 3, 3, 3, 35, 35, 35, …).

The operators of the ring amount to pointwise addition and multiplication of such sequences. This is well defined because addition and multiplication commute with the "mod" operator.

Moreover, every sequence (a_n) where the first element is not 0 has an inverse. In that case, for every n, a_n and p are coprime, and so a_n and p^n are relatively prime. Therefore, each a_n has an inverse mod p^n, and the sequence of these inverses, (b_n), is the sought inverse of (a_n). For example, consider the p-adic integer corresponding to the natural number 7; as a 2-adic number, it would be written (1, 3, 7, 7, 7, 7, 7, ...). This object's inverse would be written as an ever-increasing sequence that begins (1, 3, 7, 7, 23, 55, 55, 183, 439, 439, 1463 ...). Naturally, this 2-adic integer has no corresponding natural number.

Every such sequence can alternatively be written as a series. For instance, in the 3-adics, the sequence (2, 8, 8, 35, 35, ...) can be written as $2 + 2\cdot3 + 0\cdot3^2 + 1\cdot3^3 + 0\cdot3^4 + \ldots$ The partial sums of this latter series are the elements of the given sequence.

The ring of p-adic integers has no zero divisors, so we can take the field of fractions to get the field $\mathbf{Q}_p$ of p-adic numbers. Note that in this field of fractions, every non-integer p-adic number can be uniquely written as $p^{-n}u$ with a natural number n and a unit in the p-adic integers u. This means that

$$\mathbf{Q}_p = \mathrm{Quot}(\mathbf{Z}_p) \cong (p^{\mathbf{N}})^{-1}\mathbf{Z}_p.$$

Note that $S^{-1}A$, where $S = p^{\mathbf{N}} = \{p^n : n \in \mathbf{N}\}$ is a multiplicative subset (contains the unit and closed under multiplication) of a commutative ring with unit A, is an algebraic construction called the **ring of fractions** or localization of A by S.

Properties

Cardinality

$\mathbf{Z}_p$ is the inverse limit of the finite rings $\mathbf{Z}/p^k\mathbf{Z}$, which is uncountable—in fact, has the cardinality of the continuum. Accordingly, the field $\mathbf{Q}_p$ is uncountable. The endomorphism ring of the Prüfer p-group of rank n, denoted $\mathbf{Z}(p^\infty)^n$, is the ring of n × n matrices over $\mathbf{Z}_p$; this is sometimes referred to as the Tate module.

Topology

Define a topology on $\mathbf{Z}_p$ by taking as a basis of open sets all sets of the form

$U_a(n) = \{n + p^a : \in \mathbf{Z}_p\}$.

where a is a non-negative integer and n is an integer in $[1, p^a]$. For example, in the dyadic integers, $U_1(1)$ is the set of odd numbers. $U_a(n)$ is the set of all p-adic integers whose difference from n has p-adic absolute value less than p^{1-a}. Then $\mathbf{Z}_p$ is a compactification of **Z**, under the derived topology (it is not a compactification of **Z** with its usual discrete topology). The relative topology on **Z** as a subset of $\mathbf{Z}_p$ is called the p-adic topology on **Z**.

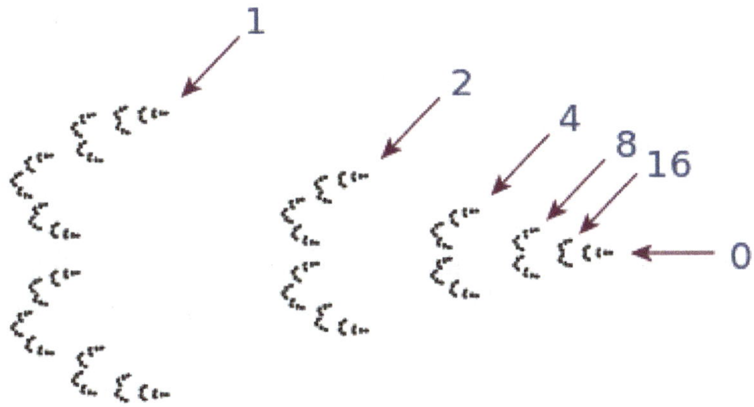

A scheme showing the topology of the dyadic (or indeed p-adic) integers. Each clump is an open set made up of other clumps. The numbers in the left-most quarter (containing 1) are all the odd numbers. The next group to the right is the even numbers not divisible by 4.

The topology of $\mathbf{Z}_p$ is that of a Cantor set C. For instance, we can make a continuous 1-to-1 mapping between the dyadic integers and the Cantor set expressed in base 3 by mapping $\ldots d_2 d_1 d_0$ in $\mathbf{Z}_2$ to $0.e_0 e_1 e_2 \ldots_3$ in C, where $e_n = 2d_n$. Using a different mapping, in which the integers go to just part of the Cantor set, one can show that the topology of $\mathbf{Q}_p$ is that of a Cantor set minus a point (such as the right-most point). In particular, $\mathbf{Z}_p$ is compact while $\mathbf{Q}_p$ is not; it is only locally compact. As metric spaces, both $\mathbf{Z}_p$ and $\mathbf{Q}_p$ are complete.

Metric Completions and Algebraic Closures

$\mathbf{Q}_p$ contains **Q** and is a field of characteristic 0. This field cannot be turned into an ordered field.

R has only a single proper algebraic extension: **C**; in other words, this quadratic extension is already algebraically closed. By contrast, the algebraic closure of $\mathbf{Q}_p$, denoted $\mathbf{Q}_p$, has infinite degree, i.e. $\mathbf{Q}_p$ has infinitely many inequivalent algebraic extensions. Also contrasting the case of real numbers, although there is a unique extension of the p-adic valuation to $\mathbf{Q}_p$, the latter is not (metrically) complete. Its (metric) completion is called $\mathbf{C}_p$ or Ω_p. Here an end is reached, as $\mathbf{C}_p$ is algebraically closed. However unlike **C** this field is not locally compact.

$\mathbf{C}_p$ and **C** are isomorphic as rings, so we may regard $\mathbf{C}_p$ as **C** endowed with an exotic metric. It should be noted that the proof of existence of such a field isomorphism relies on the axiom of choice, and does not provide an explicit example of such an isomorphism.

If **K** is a finite Galois extension of $\mathbf{Q}_p$, the Galois group Gal($\mathbf{K}/\mathbf{Q}_p$) is solvable. Thus, the Galois group Gal($\mathbf{Q}_p/\mathbf{Q}_p$) is prosolvable.

Multiplicative Group of Q_p

$\mathbf{Q}_p$ contains the n-th cyclotomic field (n > 2) if and only if n | p − 1. For instance, the n-th cyclotomic field is a subfield of $\mathbf{Q}_{13}$ if and only if n = 1, 2, 3, 4, 6, or 12. In particular, there is no multiplicative p-torsion in $\mathbf{Q}_p$, if p > 2. Also, −1 is the only non-trivial torsion element in $\mathbf{Q}_2$.

Given a natural number k, the index of the multiplicative group of the k-th powers of the non-zero elements of $\mathbf{Q}_p$ in $\mathbf{Q}\times p$ is finite.

The number e, defined as the sum of reciprocals of factorials, is not a member of any p-adic field; but $e^p \in \mathbf{Q}_p$ ($p \neq 2$). For p = 2 one must take at least the fourth power. (Thus a number with similar properties as e – namely a p-th root of e^p – is a member of $\mathbf{Q}_p$ for all p.)

Analysis on Q_p

The only real functions whose derivative is zero are the constant functions. This is not true over $\mathbf{Q}_p$. For instance, the function

$$\begin{cases} f : \mathbf{Q}_p \to \mathbf{Q}_p \\ f(x) = \begin{cases} |x|_p^{-2} & x \neq 0 \\ 0 & x = 0 \end{cases} \end{cases}$$

has zero derivative everywhere but is not even locally constant at 0.

If we let **R** be denoted $\mathbf{Q}_\infty$, then given any elements r_∞, r_2, r_3, r_5, r_7, ... where $r_p \in \mathbf{Q}_p$, it is possible to find a sequence (x_n) in **Q** such that for all p (including ∞), the limit of x_n in $\mathbf{Q}_p$is r_p.

Rational Arithmetic

Eric Hehner and Nigel Horspool proposed in 1979 the use of a p-adic representation for rational numbers on computers called Quote notation. The primary advantage of such a representation is that addition, subtraction, and multiplication can be done in a straightforward manner analogous to similar methods for binary integers; and division is even simpler, resembling multiplication. However, it has the disadvantage that representations can be much larger than simply storing the numerator and denominator in binary; for example, if $2^n - 1$ is a Mersenne prime, its reciprocal will require $2^n - 1$ bits to represent.

Generalizations and Related Concepts

The reals and the p-adic numbers are the completions of the rationals; it is also possible to complete other fields, for instance general algebraic number fields, in an analogous way. This will be described now.

Suppose D is a Dedekind domain and E is its field of fractions. Pick a non-zero prime ideal P of D. If x is a non-zero element of E, then xD is a fractional ideal and can be uniquely factored as a product of positive and negative powers of non-zero prime ideals of D. We write $\mathrm{ord}_P(x)$ for the exponent of P in this factorization, and for any choice of number c greater than 1 we can set

$$|x|_P = c^{-\mathrm{ord}_P(x)}.$$

Completing with respect to this absolute value $|.|_P$ yields a field E_P, the proper generalization of the field of p-adic numbers to this setting. The choice of c does not change the completion (different choices yield the same concept of Cauchy sequence, so the same completion). It is convenient, when the residue field D/P is finite, to take for c the size of D/P.

For example, when E is a number field, Ostrowski's theorem says that every non-trivial non-Archimedean absolute value on E arises as some $|.|_P$. The remaining non-trivial absolute values on E arise from the different embeddings of E into the real or complex numbers. (In fact, the non-Archimedean absolute values can be considered as simply the different embeddings of E into the fields $\mathbf{C}_p$, thus putting the description of all the non-trivial absolute values of a number field on a common footing.)

Often, one needs to simultaneously keep track of all the above-mentioned completions when E is a number field (or more generally a global field), which are seen as encoding "local" information. This is accomplished by adele rings and idele groups.

Local–Global Principle

Helmut Hasse's local–global principle is said to hold for an equation if it can be solved over the rational numbers if and only if it can be solved over the real numbers and over the p-adic numbers for every prime p. This principle holds e.g. for equations given by quadratic forms, but fails for higher polynomials in several indeterminates.

REFERENCES

1. Gilsdorf, Thomas E. Introduction to Cultural Mathematics: With Case Studies in the Otomies and Incas, John Wiley & Sons, Feb 24, 2012.
2. Restivo, S. Mathematics in Society and History, Springer Science & Business Media, Nov 30, 1992.
3. Ore, Oystein. Number Theory and Its History, Courier Dover Publications.
4. Gouvea, Fernando Q. The Princeton Companion to Mathematics, Chapter II.1, "The Origins of Modern Mathematics", p. 82. Princeton University Press, September 28, 2008. ISBN 978-0691118802.
5. Weisstein, Eric W. "Natural Number". MathWorld.
6. "natural number", Merriam-Webster.com, Merriam-Webster, retrieved 4 October

2014

7. "Base System". www.chalkstreet.com. 29 June 2016. Retrieved 31 August 2016.
8. Suppes, Patrick (1972). Axiomatic Set Theory. Courier Dover Publications. p. 1. ISBN 0-486-61630-4.
9. Weisstein, Eric W. "Integer". MathWorld.
10. Russell, Bertrand (1919). Introduction to Mathematical Philosophy. Routledge. p. 56. ISBN 0-415-09604-9.
11. Marshak, A., The Roots of Civilisation; Cognitive Beginnings of Man's First Art, Symbol and Notation, (Weidenfeld & Nicolson, London: 1972), 81ff.
12. "Egyptian Mathematical Papyri – Mathematicians of the African Diaspora". Math.buffalo.edu. Retrieved 2012-01-30.
13. "Historia Matematica Mailing List Archive: Re: [HM] The Zero Story: a question". Sunsite.utk.edu. 1999-04-26. Retrieved 2012-01-30.
14. Staszkow, Ronald; Robert Bradshaw (2004). The Mathematical Palette (3rd ed.). Brooks Cole. p. 41. ISBN 0-534-40365-4.
15. Smith, David Eugene (1958). History of Modern Mathematics. Dover Publications. p. 259. ISBN 0-486-20429-4.
16. Selin, Helaine, ed. (2000). Mathematics across cultures: the history of non-Western mathematics. Kluwer Academic Publishers. p. 451. ISBN 0-7923-6481-3.
17. Bogomolny, A. "What's a number?". Interactive Mathematics Miscellany and Puzzles. Retrieved 11 July 2010.
18. Tobias Dantzig, Number, the language of science; a critical survey written for the cultured non-mathematician, New York, The Macmillan company, 1930.
19. Erich Friedman, What's special about this number?
20. Steven Galovich, Introduction to Mathematical Structures, Harcourt Brace Javanovich, 23 January 1989, ISBN 0-15-543468-3.
21. Paul Halmos, Naive Set Theory, Springer, 1974, ISBN 0-387-90092-6.
22. Morris Kline, Mathematical Thought from Ancient to Modern Times, Oxford University Press, 1972.
23. Alfred North Whitehead and Bertrand Russell, Principia Mathematica to *56, Cambridge University Press, 1910.
24. George I. Sanchez, Arithmetic in Maya, Austin-Texas, 1961.

CHAPTER 2

Number Theory: An overview

Chapter Contents

- **Introduction**
- **History**
- **Main Subdivisions**
- **Recent Approaches and Subfields**
- **Applications**

Number theory, also known as higher arithmetic, is a branch of mathematics concerned with the properties of integer s, rational number s, irrational number s, and real number s. Sometimes the discipline is considered to include the imaginary and complex numbers as well.

Formally, numbers are represented in terms of set s; there are various schemes for doing this. However, there are other ways to represent numbers -- for example, as angles, as points on a line, as points on a plane, or as points in space.

The integers and rational numbers can be symbolized and completely defined by numeral s. The system of numeration commonly used today was developed from systems used in Arab texts, although some scholars believe they were first used in India. The so-called Arabic numerals are 0, 1, 2, 3, 4, 5, 6, 7, 8, and 9.

INTRODUCTION

Number theory or, in older usage, arithmetic is a branch of pure mathematics devoted primarily to the study of the integers. It is sometimes called "The Queen of Mathematics" because of its foundational place in the discipline. Number theorists study prime numbers as well as the properties of objects made out of integers (e.g., rational numbers) or defined as generalizations of the integers (e.g., algebraic integers). Integers can be considered either in themselves or as solutions to equations (Diophantine geometry). Questions in number theory are often best understood through the study of analytical objects (e.g., the Riemann zeta function) that encode properties of the integers, primes or other number-theoretic objects in some fashion (analytic number theory). One may also study real numbers in relation to rational numbers, e.g., as approximated by the latter (Diophantine approximation).

The older term for number theory is *arithmetic*. By the early twentieth century, it had been superseded by "number theory". (The word "arithmetic" is used by the general public to mean "elementary calculations"; it has also acquired other meanings in mathematical logic, as in *Peano arithmetic*, and computer science, as in *floating point arithmetic*.) The use of the term *arithmetic* for *number theory* regained some ground in the second half of the 20th century, arguably in part due to French influence. In particular, *arithmetical* is preferred as an adjective to *number-theoretic*.

A Lehmer sieve, which is a primitive digital computer once used for finding primes and solving simple Diophantine equations.

HISTORY

Origins

Dawn of Arithmetic

The first historical find of an arithmetical nature is a fragment of a table: the broken clay tablet Plimpton 322 (Larsa, Mesopotamia, ca. 1800 BCE) contains a list of "Pythagorean triples", i.e., integers (a, b, c) such that $a^2 + b^2 = c^2$. The triples are too many and too large to have been obtained by brute force. The heading over the first column reads: "The *takiltum* of the diagonal which has been subtracted such that the width..."

The Plimpton 322 tablet

The table's layout suggests that it was constructed by means of what amounts, in modern language, to the identity

$$\left(\frac{1}{2}\left(x-\frac{1}{x}\right)\right)^2+1=\left(\frac{1}{2}\left(x+\frac{1}{x}\right)\right)^2,$$

which is implicit in routine Old Babylonian exercises. If some other method was used, the triples were first constructed and then reordered by c/a, presumably for actual use as a "table", i.e., with a view to applications.

It is not known what these applications may have been, or whether there could have been any; Babylonian astronomy, for example, truly flowered only later. It has been suggested instead that the table was a source of numerical examples for school problems.

While Babylonian number theory—or what survives of Babylonian mathematics that can be called thus—consists of this single, striking fragment, Babylonian algebra (in the secondary-school sense of "algebra") was exceptionally well developed. Late Neoplatonic sources state that Pythagoras learned mathematics from the Babylonians. Much earlier sources state that Thales and Pythagoras traveled and studied in Egypt.

Euclid IX 21—34 is very probably Pythagorean; it is very simple material ("odd times even is even", "if an odd number measures [= divides] an even number, then it also measures [= divides] half of it"), but it is all that is needed to prove that $\sqrt{2}$ is irrational. Pythagorean mystics gave great importance to the odd and the even. The discovery that $\sqrt{2}$ is irrational is credited to the early Pythagoreans (pre-Theodorus). By revealing (in modern terms) that numbers could be irrational, this discovery seems to have provoked the first foundational crisis in mathematical history; its proof or its divulgation are sometimes credited to Hippasus, who was expelled or split from the Pythagorean sect. This forced a distinction between *numbers* (integers and the rationals—the subjects of arithmetic), on the one hand, and *lengths* and *proportions* (which we would identify with real numbers, whether rational or not), on the other hand.

The Pythagorean tradition spoke also of so-called polygonal or figurate numbers. While square numbers, cubic numbers, etc., are seen now as more natural than triangular numbers, pentagonal numbers, etc., the study of the sums of triangular and pentagonal numbers would prove fruitful in the early modern period (17th to early 19th century).

We know of no clearly arithmetical material in ancient Egyptian or Vedic sources, though there is some algebra in both. The Chinese remainder theorem appears as an exercise in Sun Zi's *Suan Ching*, also known as *The Mathematical Classic of Sun Zi* (3rd, 4th or 5th century CE.) (There is one important step glossed over in Sun Zi's solution: it is the problem that was later solved by Āryabhaa's Kuaka – see below.)

There is also some numerical mysticism in Chinese mathematics, but, unlike that of the Pythagoreans, it seems to have led nowhere. Like the Pythagoreans' perfect numbers, magic squares have passed from superstition into recreation.

Classical Greece and the early Hellenistic period

Aside from a few fragments, the mathematics of Classical Greece is known to us either through the reports of contemporary non-mathematicians or through mathematical works from the early Hellenistic period. In the case of number theory, this means, by and large, *Plato* and *Euclid*, respectively.

Plato had a keen interest in mathematics, and distinguished clearly between arithmetic and calculation. (By *arithmetic* he meant, in part, theorizing on number, rather than what *arithmetic* or *number theory* have come to mean.) It is through one of Plato's dialogues—namely, *Theaetetus*—that we know that Theodorus had proven that $\sqrt{3}, \sqrt{5}, \ldots, \sqrt{17}$ are irrational. Theaetetus was, like Plato, a disciple of Theodorus's; he worked on distinguishing different kinds of incommensurables, and was thus arguably a pioneer in the study of number systems. (Book X of Euclid's Elements is described by Pappus as being largely based on Theaetetus's work.)

Euclid devoted part of his *Elements* to prime numbers and divisibility, topics that belong unambiguously to number theory and are basic to it (Books VII to IX of Euclid's Elements). In particular, he gave an algorithm for computing the greatest common divisor of two numbers (the Euclidean algorithm; *Elements*, Prop. VII.2) and the first known proof of the infinitude of primes (*Elements*, Prop. IX.20).

In 1773, Lessing published an epigram he had found in a manuscript during his work as a librarian; it claimed to be a letter sent by Archimedes to Eratosthenes. The epigram proposed what has become known as Archimedes' cattle problem; its solution (absent from the manuscript) requires solving an indeterminate quadratic equation (which reduces to what would later be misnamed Pell's equation). As far as we know, such equations were first successfully treated by the Indian school. It is not known whether Archimedes himself had a method of solution.

Diophantus

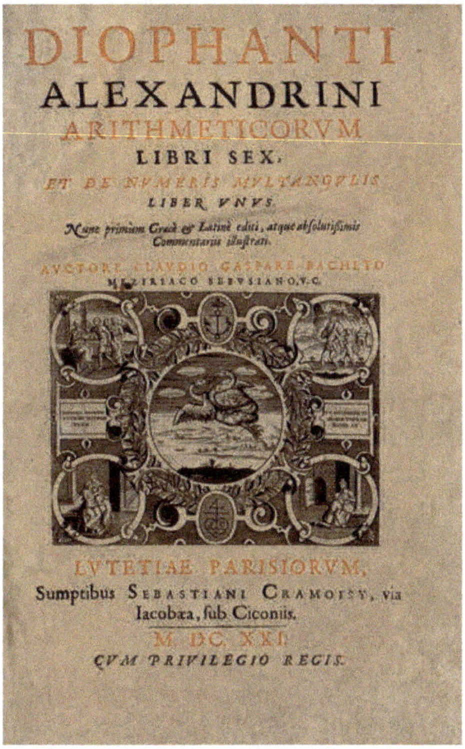
DIOPHANTI
ALEXANDRINI
ARITHMETICORVM
LIBRI SEX,
ET DE NVMERIS MVLTANGVLIS
LIBER VNVS.
Nunc primùm Græcè & Latinè editi, atque absolutissimis Commentariis illustrati.
AVCTORE CLAVDIO GASPARE BACHETO
MEZIRIACO SEBVSIANO, V.C.
LVTETIAE PARISIORVM,
Sumptibus SEBASTIANI CRAMOISY, via Iacobæa, sub Ciconiis.
M. DC. XXI.
CVM PRIVILEGIO REGIS.

Title page of the 1621 edition of Diophantus' *Arithmetica*, translated into Latin by Claude Gaspard Bachet de Méziriac.

Very little is known about Diophantus of Alexandria; he probably lived in the third century CE, that is, about five hundred years after Euclid. Six out of the thirteen books of Diophantus's *Arithmetica* survive in the original Greek; four more books survive in an Arabic translation. The *Arithmetica* is a collection of worked-out problems where the task is invariably to find rational solutions to a system of polynomial equations, usually of the form $f(x, y) = z^2$ or $f(x, y, z) = w^2$. Thus, nowadays, we speak of *Diophantine equations* when we speak of polynomial equations to which rational or integer solutions must be found.

One may say that Diophantus was studying rational points — i.e., points whose coordinates are rational — on curves and algebraic varieties; however, unlike the Greeks of the Classical period, who did what we would now call basic algebra in geometrical terms, Diophantus did what we would now call basic algebraic geometry in purely algebraic terms. In modern language, what Diophantus did was to find rational parametrizations of varieties; that is, given an equation of the form (say) $f(x_1, x_2, x_3) = 0$, his aim was to find (in essence) three rational functions g_1, g_2, g_3 such that, for all values of r and s, setting $x_i = g_i(r, s)$ gives a solution to $i = 1, 2, 3$ gives a solution to $f(x_1, x_2, x_3) = 0.$

Diophantus also studied the equations of some non-rational curves, for which no rational parametrisation is possible. He managed to find some rational points on these curves (elliptic curves, as it happens, in what seems to be their first known occurrence) by means of what amounts to a tangent construction: translated into coordinate geometry (which did not exist in Diophantus's time), his method would be visualised as drawing a tangent to a curve at a known rational point, and then finding the other point of intersection of the tangent with the curve; that other point is a new rational point. (Diophantus also resorted to what could be called a special case of a secant construction.)

While Diophantus was concerned largely with rational solutions, he assumed some results on integer numbers, in particular that every integer is the sum of four squares (though he never stated as much explicitly).

Āryabhaa, Brahmagupta, Bhāskara

While Greek astronomy probably influenced Indian learning, to the point of introducing trigonometry, it seems to be the case that Indian mathematics is otherwise an indigenous tradition;in particular, there is no evidence that Euclid's Elements reached India before the 18th century.

Āryabhaa (476–550 CE) showed that pairs of simultaneous congruences $n \equiv a_1 \bmod m_1$, $n \equiv a_2 \bmod m_2$ could be solved by a method he called *kuaka*, or *pulveriser*; this is a procedure close to (a generalisation of) the Euclidean algorithm, which was probably discovered independently in India. Āryabhaa seems to have had in mind applications to astronomical calculations.

Brahmagupta (628 CE) started the systematic study of indefinite quadratic equations—in particular, the misnamed Pell equation, in which Archimedes may have first been interested, and which did not start to be solved in the West until the time of Fermat and Euler. Later Sanskrit authors would follow, using Brahmagupta's technical terminology. A general procedure (the chakravala, or "cyclic method") for solving Pell's equation was finally found by Jayadeva (cited in the eleventh century; his work is otherwise lost); the earliest surviving exposition appears in Bhāskara II's Bīja-gaita (twelfth century).

Indian mathematics remained largely unknown in Europe until the late eighteenth century; Brahmagupta and Bhāskara's work was translated into English in 1817 by Henry Colebrooke.

Arithmetic in the Islamic golden age

In the early ninth century, the caliph Al-Ma'mun ordered translations of many Greek mathematical works and at least one Sanskrit work (the *Sindhind*, which may or may not be Brahmagupta's Brāhmasphuṭasiddhānta). Diophantus's main work, the *Arithmetica*, was translated into Arabic by Qusta ibn Luqa (820–912). Part of the treatise *al-Fakhri* (by al-Karajī, 953 – ca. 1029) builds on it to some extent. According to Rashed Roshdi, Al-Karajī's contemporary Ibn al-Haytham knew what would later be called Wilson's theorem.

Al-Haytham seen by the West: frontispice of *Selenographia*, showing Alhasen [*sic*] representing knowledge through reason, and Galileo representing knowledge through the senses.

Western Europe in the Middle Ages

Other than a treatise on squares in arithmetic progression by Fibonacci — who lived and studied in north Africa and Constantinople during his formative years, ca. 1175–1200 — no number theory to speak of was done in western Europe during the Middle Ages. Matters started to change in Europe in the late Renaissance, thanks to a renewed study of the works of Greek antiquity. A catalyst was the textual emendation and translation into Latin of Diophantus's *Arithmetica* (Bachet, 1621, following a first attempt by Xylander, 1575).

Early Modern Number Theory

Fermat

Pierre de Fermat (1601–1665) never published his writings; in particular, his work on number theory is contained almost entirely in letters to mathematicians and in private marginal notes. He wrote down nearly no proofs in number theory; he had no models in the area. He did make repeated use of mathematical induction, introducing the method of infinite descent.

Pierre de Fermat

One of Fermat's first interests was perfect numbers (which appear in Euclid, *Elements* IX) and amicable numbers; this led him to work on integer divisors, which were from the beginning among the subjects of the correspondence (1636 onwards) that put him in touch with the mathematical community of the day. He had already studied Bachet's edition of Diophantus carefully; by 1643, his interests had shifted largely to Diophantine problems and sums of squares (also treated by Diophantus).

Fermat's achievements in arithmetic include:

- Fermat's little theorem (1640), stating that, if a is not divisible by a prime p, then $a^{p-1} \equiv 1 \bmod p$.
- If a and b are coprime, then $a^2 + b^2$ is not divisible by any prime congruent to −1 modulo 4; and every prime congruent to 1 modulo 4 can be written in the form $a^2 + b^2$. These two statements also date from 1640; in 1659, Fermat stated to Huygens that he had proven the latter statement by the method of infinite descent. Fermat and Frenicle also did some work (some of it erroneous) on other quadratic forms.
- Fermat posed the problem of solving $x^2 - Ny^2 = 1$ as a challenge to English mathematicians (1657). The problem was solved in a few months by Wallis and Brouncker. Fermat considered their solution valid, but pointed out they had provided an algorithm without a proof (as had Jayadeva and Bhaskara, though Fermat would never know this.) He states that a proof can be found by descent.
- Fermat developed methods for (doing what in our terms amounts to) finding points on curves of genus 0 and 1. As in Diophantus, there are many special

procedures and what amounts to a tangent construction, but no use of a secant construction.

- Fermat states and proves (by descent) in the appendix to *Observations on Diophantus* (Obs. XLV) that $x^4 + y^4 = z^4$ has no non-trivial solutions in the integers. Fermat also mentioned to his correspondents that $x^3 + y^3 = z^3$ has no non-trivial solutions, and that this could be proven by descent. The first known proof is due to Euler (1753; indeed by descent).

Fermat's claim ("Fermat's last theorem") to have shown there are no solutions to $x^n + y^n = z^n$ for all $n \geq 3$ (the only known proof of which is beyond his methods) appears only in his annotations on the margin of his copy of Diophantus; he never claimed this to others and thus would have had no need to retract it if he found any mistake in his supposed proof.

Euler

Leonhard Euler

The interest of Leonhard Euler (1707–1783) in number theory was first spurred in 1729, when a friend of his, the amateur Goldbach, pointed him towards some of Fermat's work on the subject. This has been called the "rebirth" of modern number theory, after Fermat's relative lack of success in getting his contemporaries' attention for the subject. Euler's work on number theory includes the following:

- *Proofs for Fermat's statements.* This includes Fermat's little theorem (generalised by Euler to non-prime moduli); the fact that $p = x^2 + y^2$ if and only if $p \equiv 1 \bmod 4$; initial work towards a proof that every integer is the sum of four squares (the first complete proof is by Joseph-Louis Lagrange (1770), soon improved by Euler himself); the lack of non-zero integer solutions to $x^4 + y^4 = z^2$ (implying the case *n=4* of

Fermat's last theorem, the case *n=3* of which Euler also proved by a related method).

- *Pell's equation*, first misnamed by Euler. He wrote on the link between continued fractions and Pell's equation.
- *First steps towards analytic number theory.* In his work of sums of four squares, partitions, pentagonal numbers, and the distribution of prime numbers, Euler pioneered the use of what can be seen as analysis (in particular, infinite series) in number theory. Since he lived before the development of complex analysis, most of his work is restricted to the formal manipulation of power series. He did, however, do some very notable (though not fully rigorous) early work on what would later be called the Riemann zeta function.
- *Quadratic forms*. Following Fermat's lead, Euler did further research on the question of which primes can be expressed in the form $x^2 + Ny^2$, some of it prefiguring quadratic reciprocity.
- *Diophantine equations*. Euler worked on some Diophantine equations of genus 0 and 1. In particular, he studied Diophantus's work; he tried to systematise it, but the time was not yet ripe for such an endeavour – algebraic geometry was still in its infancy. He did notice there was a connection between Diophantine problems and elliptic integrals, whose study he had himself initiated.

Lagrange, Legendre and Gauss

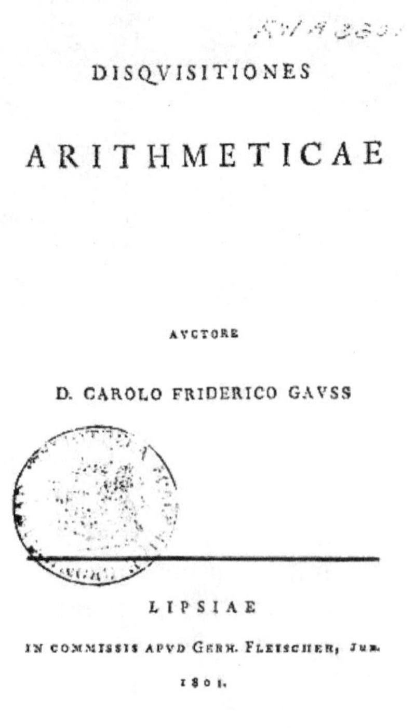
DISQVISITIONES

ARITHMETICAE

AVCTORE

D. CAROLO FRIDERICO GAVSS

LIPSIAE

IN COMMISSIS APVD GERH. FLEISCHER, IVN.

1801.

Carl Friedrich Gauss's Disquisitiones Arithmeticae, first edition

Joseph-Louis Lagrange (1736–1813) was the first to give full proofs of some of Fermat's and Euler's work and observations – for instance, the four-square theorem and the basic theory of the misnamed "Pell's equation" (for which an algorithmic solution was found by Fermat and his contemporaries, and also by Jayadeva and Bhaskara II before them.) He also studied quadratic forms in full generality (as opposed to $mX^2 + nY^2$) — defining their equivalence relation, showing how to put them in reduced form, etc.

Adrien-Marie Legendre (1752–1833) was the first to state the law of quadratic reciprocity. He also conjectured what amounts to the prime number theorem and Dirichlet's theorem on arithmetic progressions. He gave a full treatment of the equation $ax^2 + by^2 + cz^2 = 0$ and worked on quadratic forms along the lines later developed fully by Gauss. In his old age, he was the first to prove "Fermat's last theorem" for n=5 (completing work by Peter Gustav Lejeune Dirichlet, and crediting both him and Sophie Germain).

Carl Friedrich Gauss

In his *Disquisitiones Arithmeticae* (1798), Carl Friedrich Gauss (1777–1855) proved the law of quadratic reciprocity and developed the theory of quadratic forms (in particular, defining their composition). He also introduced some basic notation (congruences) and devoted a section to computational matters, including primality tests. The last section of the *Disquisitiones* established a link between roots of unity and number theory:

The theory of the division of the circle...which is treated in sec. 7 does not belong by itself to arithmetic, but its principles can only be drawn from higher arithmetic.

In this way, Gauss arguably made a first foray towards both Évariste Galois's work and algebraic number theory.

Maturity and Division into Subfields

Ernst Kummer

Peter Gustav Lejeune Dirichlet

Starting early in the nineteenth century, the following developments gradually took place:

- The rise to self-consciousness of number theory (or *higher arithmetic*) as a field of study.
- The development of much of modern mathematics necessary for basic modern number theory: complex analysis, group theory, Galois theory—accompanied by greater rigor in analysis and abstraction in algebra.
- The rough subdivision of number theory into its modern subfields—in particular, analytic and algebraic number theory.

Algebraic number theory may be said to start with the study of reciprocity and cyclotomy, but truly came into its own with the development of abstract algebra and early ideal theory and valuation theory; see below. A conventional starting point for analytic number theory is Dirichlet's theorem on arithmetic progressions (1837), whose proof introduced L-functions and involved some asymptotic analysis and a limiting process on a real variable. The first use of analytic ideas in number theory actually goes back to Euler (1730s), who used formal power series and non-rigorous (or implicit) limiting arguments. The use of *complex* analysis in number theory comes later: the work of Bernhard Riemann (1859) on the zeta function is the canonical starting point; Jacobi's four-square theorem (1839), which predates it, belongs to an initially different strand that has by now taken a leading role in analytic number theory (modular forms).

The history of each subfield is briefly addressed in its own section below; see the main section of each subfield for fuller treatments. Many of the most interesting questions in each area remain open and are being actively worked on.

MAIN SUBDIVISIONS

Elementary Tools

The term *elementary* generally denotes a method that does not use complex analysis. For example, the prime number theorem was first proven using complex analysis in 1896, but an elementary proof was found only in 1949 by Erdős and Selberg. The term is somewhat ambiguous: for example, proofs based on complex Tauberian theorems (e.g. Wiener–Ikehara) are often seen as quite enlightening but not elementary, in spite of using Fourier analysis, rather than complex analysis as such. Here as elsewhere, an *elementary* proof may be longer and more difficult for most readers than a non-elementary one.

Number theory has the reputation of being a field many of whose results can be stated to the layperson. At the same time, the proofs of these results are not particularly accessible, in part because the range of tools they use is, if anything, unusually broad within mathematics.

Analytic Number Theory

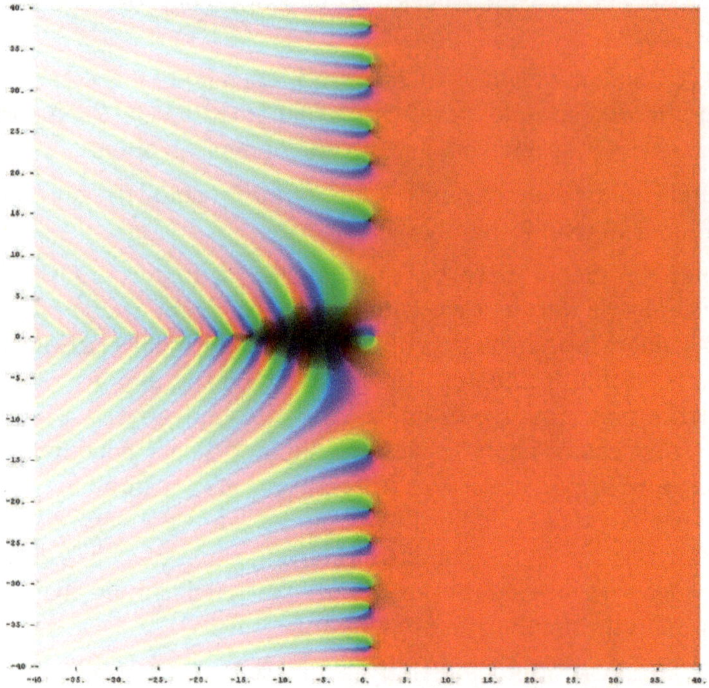

Riemann zeta function $\zeta(s)$ in the complex plane. The color of a point s gives the value of $\zeta(s)$: dark colors denote values close to zero and hue gives the value's argument.

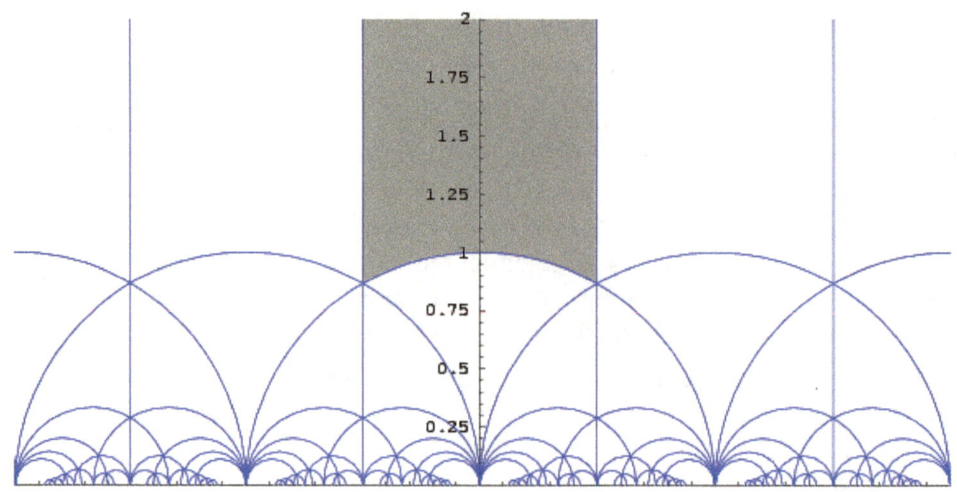

The action of the modular group on the upper half plane. The region in grey is the standard fundamental domain.

Analytic number theory ***may be defined***

- in terms of its tools, as the study of the integers by means of tools from real and complex analysis; or
- in terms of its concerns, as the study within number theory of estimates on size and density, as opposed to identities.

Some subjects generally considered to be part of analytic number theory, e.g., sieve theory, are better covered by the second rather than the first definition: some of sieve theory, for instance, uses little analysis, yet it does belong to analytic number theory.

The following are examples of problems in analytic number theory: the prime number theorem, the Goldbach conjecture (or the twin prime conjecture, or the Hardy–Littlewood conjectures), the Waring problem and the Riemann hypothesis. Some of the most important tools of analytic number theory are the circle method, sieve methods and L-functions (or, rather, the study of their properties). The theory of modular forms (and, more generally, automorphic forms) also occupies an increasingly central place in the toolbox of analytic number theory.

One may ask analytic questions about algebraic numbers, and use analytic means to answer such questions; it is thus that algebraic and analytic number theory intersect. For example, one may define prime ideals (generalizations of prime numbers in the field of algebraic numbers) and ask how many prime ideals there are up to a certain size. This question can be answered by means of an examination of Dedekind zeta functions, which are generalizations of the Riemann zeta function, a key analytic object at the roots of the subject. This is an example of a general procedure in analytic number theory: deriving information about the distribution of a sequence (here, prime ideals or prime numbers) from the analytic behavior of an appropriately constructed complex-valued function.

Algebraic Number Theory

An *algebraic number* is any complex number that is a solution to some polynomial equation $f(x) = 0$ with rational coefficients; for example, every solution x of $x^5 + (11/2)x^3 - 7x^2 + 9 = 0$ (say) is an algebraic number. Fields of algebraic numbers are also called *algebraic number fields*, or shortly *number fields*. Algebraic number theory studies algebraic number fields. Thus, analytic and algebraic number theory can and do overlap: the former is defined by its methods, the latter by its objects of study.

It could be argued that the simplest kind of number fields (viz., quadratic fields) were already studied by Gauss, as the discussion of quadratic forms in *Disquisitiones arithmeticae* can be restated in terms of ideals and norms in quadratic fields. (A *quadratic field* consists of all numbers of the form $a + b\sqrt{d}$, where a and b are rational numbers and d is a fixed rational number whose square root is not rational.) For that matter, the 11th-century chakravala method amounts—in modern terms—to an algorithm for finding the units of a real quadratic number field. However, neither Bhāskara nor Gauss knew of number fields as such.

The grounds of the subject as we know it were set in the late nineteenth century, when *ideal numbers*, the *theory of ideals* and *valuation theory* were developed; these are three complementary ways of dealing with the lack of unique factorisation in algebraic number fields. (For example, in the field generated by the rationals and $\sqrt{-5}$, the number 6 can be factorized both as $6 = 2 \cdot 3$ and $6 = (1 + \sqrt{-5})(1 - \sqrt{-5})$; all of $2, 3, 1 + \sqrt{-5}$ and $1 - \sqrt{-5}$ are irreducible, and thus, in a naïve sense, analogous to primes among the integers.) The initial impetus for the development of ideal numbers (by Kummer) seems to have come from the study of higher reciprocity laws,i.e., generalisations of quadratic reciprocity.

Number fields are often studied as extensions of smaller number fields: a field L is said to be an *extension* of a field K if L contains K. (For example, the complex numbers C are an extension of the reals R, and the reals R are an extension of the rationals Q.) Classifying the possible extensions of a given number field is a difficult and partially open problem. Abelian extensions—that is, extensions L of K such that the Galois group Gal(L/K) of L over K is an abelian group—are relatively well understood. Their classification was the object of the programme of class field theory, which was initiated in the late 19th century (partly by Kronecker and Eisenstein) and carried out largely in 1900—1950.

An example of an active area of research in algebraic number theory is Iwasawa theory. The Langlands program, one of the main current large-scale research plans in mathematics, is sometimes described as an attempt to generalise class field theory to non-abelian extensions of number fields.

Diophantine Geometry

The central problem of *Diophantine geometry* is to determine when a Diophantine equation has solutions, and if it does, how many. The approach taken is to think of the solutions of an equation as a geometric object.

For example, an equation in two variables defines a curve in the plane. More generally, an equation, or system of equations, in two or more variables defines a curve, a surface or some other such object in n-dimensional space. In Diophantine geometry, one asks whether there are any *rational points* (points all of whose coordinates are rationals) or *integral points* (points all of whose coordinates are integers) on the curve or surface. If there are any such points, the next step is to ask how many there are and how they are distributed. A basic question in this direction is: are there finitely or infinitely many rational points on a given curve (or surface)? What about integer points?

An example here may be helpful. Consider the Pythagorean equation $x^2 + y^2 = 1$; we would like to study its rational solutions, i.e., its solutions (x,y) such that x and y are both rational. This is the same as asking for all integer solutions to $a^2 + b^2 = c^2$; any solution to the latter equation gives us a solution $x = a/c, y = b/c$ to the former. It is also the same as asking for all points with rational coordinates on the curve described by $x^2 + y^2 = 1$. (This curve happens to be a circle of radius 1 around the origin.)

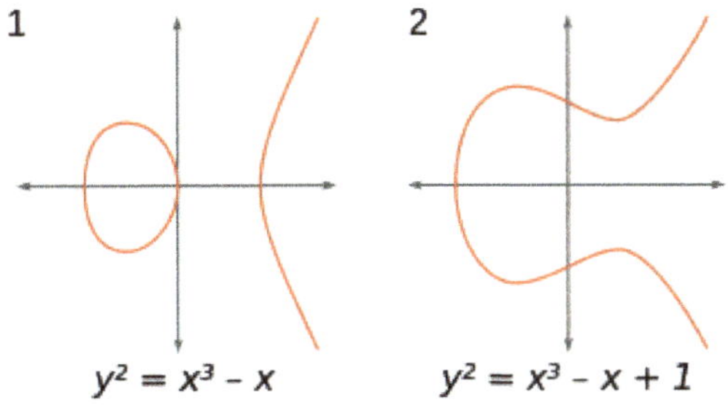

Two examples of an elliptic curve, i.e., a curve of genus 1 having at least one rational point. (Either graph can be seen as a slice of a torus in four-dimensional space.)

The rephrasing of questions on equations in terms of points on curves turns out to be felicitous. The finiteness or not of the number of rational or integer points on an algebraic curve—that is, rational or integer solutions to an equation $f(x, y) = 0$, where f is a polynomial in two variables—turns out to depend crucially on the *genus* of the curve. The *genus* can be defined as follows: allow the variables in $f(x, y) = 0$ to be complex numbers; then $f(x, y) = 0$ defines a 2-dimensional surface in (projective) 4-dimensional space (since two complex variables can be decomposed into four real variables, i.e., four dimensions). Count the number of (doughnut) holes in the surface; call this number the *genus* of $f(x, y) = 0$. Other geometrical notions turn out to be just as crucial.

There is also the closely linked area of Diophantine approximations: given a number x, how well can it be approximated by rationals? (We are looking for approximations that are good relative to the amount of space that it takes to write the rational: a/q (with $\gcd(a, q) = 1$) a good approximation to x if $|x - a/q| < \frac{1}{q^c}$, where c is large.) This question is of special interest if x is an algebraic number. If x cannot be well approximated, then some equations do not have integer or rational solutions. Moreover, several concepts (especially that of height) turn out to be crucial both in Diophantine geometry and in the study of Diophantine approximations. This question is also of special interest in transcendental number theory: if a number can be better approximated than any algebraic number, then it is a transcendental number. It is by this argument that π and e have been shown to be transcendental.

Diophantine geometry should not be confused with the geometry of numbers, which is a collection of graphical methods for answering certain questions in algebraic number theory. *Arithmetic geometry*, on the other hand, is a contemporary term for much the same domain as that covered by the term *Diophantine geometry*. The term *arithmetic geometry* is arguably used most often when one wishes to emphasise the connections to modern algebraic geometry (as in, for instance, Faltings' theorem) rather than to techniques in Diophantine approximations.

RECENT APPROACHES AND SUBFIELDS

The areas below date as such from no earlier than the mid-twentieth century, even if they are based on older material. For example, as is explained below, the matter of algorithms in number theory is very old, in some sense older than the concept of proof; at the same time, the modern study of computability dates only from the 1930s and 1940s, and computational complexity theory from the 1970s.

Probabilistic Number Theory

Take a number at random between one and a million. How likely is it to be prime? This is just another way of asking how many primes there are between one and a million. Further: how many prime divisors will it have, on average? How many divisors will it have altogether, and with what likelihood? What is the probability that it will have many more or many fewer divisors or prime divisors than the average?

Much of probabilistic number theory can be seen as an important special case of the study of variables that are almost, but not quite, mutually independent. For example, the event that a random integer between one and a million be divisible by two and the event that it be divisible by three are almost independent, but not quite.

It is sometimes said that probabilistic combinatorics uses the fact that whatever happens with probability greater than 0 must happen sometimes; one may say with equal justice that many applications of probabilistic number theory hinge on the fact that whatever is unusual must be rare. If certain algebraic objects (say, rational or integer solutions to certain equations) can be shown to be in the tail of certain sensibly defined distributions, it follows that there must be few of them; this is a very concrete non-probabilistic statement following from a probabilistic one.

At times, a non-rigorous, probabilistic approach leads to a number of heuristic algorithms and open problems, notably Cramér's conjecture.

Computations in Number Theory

While the word *algorithm* goes back only to certain readers of al-Khwārizmī, careful descriptions of methods of solution are older than proofs: such methods (that is, algorithms) are as old as any recognisable mathematics—ancient Egyptian, Babylonian, Vedic, Chinese—whereas proofs appeared only with the Greeks of the classical period. An interesting early case is that of what we now call the Euclidean algorithm. In its basic form (namely, as an algorithm for computing the greatest common divisor) it appears as Proposition 2 of Book VII in *Elements*, together with a proof of correctness. However, in the form that is often used in number theory (namely, as an algorithm for finding integer solutions to an equation $ax + by = c$, or, what is the same, for finding the quantities whose existence is assured by the Chinese remainder theorem) it first appears in the works of Āryabhaa (5th–6th century CE) as an algorithm called *kuaka* ("pulveriser"), without a proof of correctness.

There are two main questions: "can we compute this?" and "can we compute it rapidly?". Anybody can test whether a number is prime or, if it is not, split it into prime factors; doing so rapidly is another matter. We now know fast algorithms for testing primality, but, in spite of much work (both theoretical and practical), no truly fast algorithm for factoring.

The difficulty of a computation can be useful: modern protocols for encrypting messages (e.g., RSA) depend on functions that are known to all, but whose inverses (a) are known only to a chosen few, and (b) would take one too long a time to figure out on one's own. For example, these functions can be such that their inverses can be computed only if certain large integers are factorized. While many difficult computational problems outside number theory are known, most working encryption protocols nowadays are based on the difficulty of a few number-theoretical problems.

On a different note — some things may not be computable at all; in fact, this can be proven in some instances. For instance, in 1970, it was proven, as a solution to Hilbert's 10th problem, that there is no Turing machine which can solve all Diophantine equations. In particular, this means that, given a computably enumerable set of axioms, there are Diophantine equations for which there is no proof, starting from the axioms, of whether the set of equations has or does not have integer solutions. (We would necessarily be speaking of Diophantine equations for which there are no integer solutions, since, given a Diophantine equation with at least one solution, the solution itself provides a proof of the fact that a solution exists. We cannot prove, of course, that a particular Diophantine equation is of this kind, since this would imply that it has no solutions.)

APPLICATIONS

The number-theorist Leonard Dickson (1874–1954) said "Thank God that number theory is unsullied by any application". Such a view is no longer applicable to number theory. In 1974, Donald Knuth said "...virtually every theorem in elementary number theory arises in a natural, motivated way in connection with the problem of making computers do high-speed numerical calculations". Elementary number theory is taught in discrete mathematics courses for computer scientists; and, on the other hand, number theory also has applications to the continuous in numerical analysis. As well as the well-known applications to cryptography, there are also applications to many other areas of mathematics.

REFERENCES

1. Neugebauer & Sachs 1945, p. 40. The term takiltum is problematic. Robson prefers the rendering "The holding-square of the diagonal from which 1 is torn out, so that the short side comes up...".Robson 2001, p. 192
2. Robson 2001, p. 189. Other sources give the modern formula {\displaystyle (p{2}-q{2},2pq,p{2}+q{2})} (p{2}-q{2},2pq,p{2}+q{2}). Van der Waerden gives both the modern formula and what amounts to the form preferred by Robson. (van der Waerden 1961, p. 79)
3. van der Waerden 1961, p. 184.
4. Neugebauer (Neugebauer 1969, pp. 36–40) discusses the table in detail and mentions in passing Euclid's method in modern notation (Neugebauer 1969, p. 39).
5. Friberg 1981, p. 302.
6. van der Waerden 1961, p. 43.
7. Iamblichus, Life of Pythagoras,(trans. e.g. Guthrie 1987) cited in van der Waerden 1961, p. 108. See also Porphyry, Life of Pythagoras, paragraph 6, in Guthrie

1987 Van der Waerden (van der Waerden 1961, pp. 87–90) sustains the view that Thales knew Babylonian mathematics.

8. Herodotus (II. 81) and Isocrates (Busiris 28), cited in: Huffman 2011. On Thales, see Eudemus ap. Proclus, 65.7, (e.g. Morrow 1992, p. 52) cited in: O'Grady 2004, p. 1. Proclus was using a work by Eudemus of Rhodes (now lost), the Catalogue of Geometers. See also introduction, Morrow 1992, p. xxx on Proclus' reliability.
9. Becker 1936, p. 533, cited in: van der Waerden 1961, p. 108.
10. Becker 1936.
11. van der Waerden 1961, p. 109.
12. Plato, Theaetetus, p. 147 B, (e.g. Jowett 1871), cited in von Fritz 2004, p. 212: "Theodorus was writing out for us something about roots, such as the roots of three or five, showing that they are incommensurable by the unit;..." See also Spiral of Theodorus.
13. von Fritz 2004.
14. Heath 1921, p. 76.
15. Sun Zi, Suan Ching, Chapter 3, Problem 26. This can be found in Lam & Ang 2004, pp. 219–220, which contains a full translation of the Suan Ching (based on Qian 1963). See also the discussion in Lam & Ang 2004, pp. 138–140.
16. The date of the text has been narrowed down to 220–420 AD (Yan Dunjie) or 280–473 AD (Wang Ling) through internal evidence (= taxation systems assumed in the text). See Lam & Ang 2004, pp. 27–28.
17. Boyer & Merzbach 1991, p. 82.
18. Vardi 1998, p. 305–319.
19. Weil 1984, pp. 17–24.
20. Plofker 2008, p. 119.
21. Any early contact between Babylonian and Indian mathematics remains conjectural (Plofker 2008, p. 42).
22. Mumford 2010, p. 387.
23. Āryabhaa, Āryabhatīya, Chapter 2, verses 32–33, cited in: Plofker 2008, pp. 134–140. See also Clark 1930, pp. 42–50. A slightly more explicit description of the kuaka was later given in Brahmagupta, Brāhmasphuasiddhānta, XVIII, 3–5 (in Colebrooke 1817, p. 325, cited in Clark 1930, p. 42).
24. Mumford 2010, p. 388.
25. Plofker 2008, p. 194.
26. Plofker 2008, p. 283.
27. Colebrooke 1817.
28. Colebrooke 1817, p. lxv, cited in Hopkins 1990, p. 302. See also the preface in Sachau 1888 cited in Smith 1958, pp. 168
29. Pingree 1968, pp. 97–125, and Pingree 1970, pp. 103–123, cited in Plofker

2008, p. 256.

30. Rashed 1980, p. 305–321.
31. Weil 1984, pp. 45–46.
32. Weil 1984, p. 118. This was more so in number theory than in other areas (remark in Mahoney 1994, p. 284). Bachet's own proofs were "ludicrously clumsy" (Weil 1984, p. 33).
33. Mahoney 1994, pp. 48, 53–54. The initial subjects of Fermat's correspondence included divisors ("aliquot parts") and many subjects outside number theory; see the list in the letter from Fermat to Roberval, 22.IX.1636, Tannery & Henry 1891, Vol. II, pp. 72, 74, cited in Mahoney 1994, p. 54.
34. Weil 1984, pp. 1–2.
35. Weil 1984, p. 53.
36. Tannery & Henry 1891, Vol. II, p. 209, Letter XLVI from Fermat to Frenicle, 1640, cited in Weil 1984, p. 56
37. Tannery & Henry 1891, Vol. II, p. 204, cited in Weil 1984, p. 63. All of the following citations from Fermat's Varia Opera are taken from Weil 1984, Chap. II. The standard Tannery & Henry work includes a revision of Fermat's posthumous Varia Opera Mathematica originally prepared by his son (Fermat 1679).
38. Tannery & Henry 1891, Vol. II, p. 213.
39. Tannery & Henry 1891, Vol. II, p. 423.
40. Weil 1984, pp. 80, 91–92.
41. Weil 1984, p. 92.
42. Weil 1984, Ch. II, sect. XV and XVI.
43. Tannery & Henry 1891, Vol. I, pp. 340–341.
44. Weil 1984, p. 115.
45. Weil 1984, pp. 115–116.
46. Weil 1984, p. 104.
47. Weil 1984, pp. 2, 172.
48. Varadarajan 2006, p. 9.
49. Weil 1984, p. 2 and Varadarajan 2006, p. 37
50. Varadarajan 2006, p. 39 and Weil 1984, pp. 176–189
51. Weil 1984, pp. 178–179.
52. Weil 1984, p. 174. Euler was generous in giving credit to others (Varadarajan 2006, p. 14), not always correctly.
53. Weil 1984, p. 183.
54. Varadarajan 2006, pp. 45–55; see also chapter III.
55. Varadarajan 2006, pp. 44–47.
56. Weil 1984, pp. 177–179.

57. Edwards 1983, pp. 285–291.
58. Varadarajan 2006, pp. 55–56.
59. Weil 1984, pp. 179–181.
60. Weil 1984, p. 181.
61. Weil 1984, pp. 327–328.
62. Weil 1984, pp. 332–334.
63. Weil 1984, pp. 337–338.
64. Goldstein & Schappacher 2007, p. 14.
65. From the preface of Disquisitiones Arithmeticae; the translation is taken from Goldstein & Schappacher 2007, p. 16
66. See the discussion in section 5 of Goldstein & Schappacher 2007. Early signs of self-consciousness are present already in letters by Fermat: thus his remarks on what number theory is, and how "Diophantus's work [...] does not really belong to [it]" (quoted in Weil 1984, p. 25).
67. Apostol 1976, p. 7.
68. Davenport & Montgomery 2000, p. 1.
69. See the proof in Davenport & Montgomery 2000, section 1
70. Iwaniec & Kowalski 2004, p. 1.
71. Varadarajan 2006, sections 2.5, 3.1 and 6.1.
72. Granville 2008, pp. 322–348.
73. See the comment on the importance of modularity in Iwaniec & Kowalski 2004, p. 1
74. Goldfeld 2003.
75. See, e.g., the initial comment in Iwaniec & Kowalski 2004, p. 1.
76. Granville 2008, section 1: "The main difference is that in algebraic number theory [...] one typically considers questions with answers that are given by exact formulas, whereas in analytic number theory [...] one looks for good approximations."
77. See the remarks in the introduction to Iwaniec & Kowalski 2004, p. 1: "However much stronger...".
78. Granville 2008, section 3: "[Riemann] defined what we now call the Riemann zeta function [...] Riemann's deep work gave birth to our subject [...]"
79. See, e.g., Montgomery & Vaughan 2007, p. 1.
80. CITEREFMilne2014, p. 2.
81. Edwards 2000, p. 79.
82. Davis, Martin; Matiyasevich, Yuri; Robinson, Julia (1976). "Hilbert's Tenth Problem: Diophantine Equations: Positive Aspects of a Negative Solution". In Felix E. Browder. Mathematical Developments Arising from Hilbert Problems. Proceedings of Symposia in Pure Mathematics. XXVIII.2. American Mathematical

Society. pp. 323–378. ISBN 0-8218-1428-1. Zbl 0346.02026. Reprinted in The Collected Works of Julia Robinson, Solomon Feferman, editor, pp.269–378, American Mathematical Society 1996.

83. "The Unreasonable Effectiveness of Number Theory", Stefan Andrus Burr, George E. Andrews, American Mathematical Soc., 1992, ISBN 978-0-8218-5501-0

84. Computer science and its relation to mathematics" DE Knuth – The American Mathematical Monthly, 1974

85. "Applications of number theory to numerical analysis", Lo-keng Hua, Luogeng Hua, Yuan Wang, Springer-Verlag, 1981, ISBN 978-3-540-10382-0

86. "Practical applications of algebraic number theory". Mathoverflow.net. Retrieved 2012-05-18.

87. "Where is number theory used in the rest of mathematics?". Mathoverflow.net. 2008-09-23. Retrieved 2012-05-18.

88. Apostol n.d.

CHAPTER 3

Algebraic Number Theory

Chapter Contents

An algebraic number field is a finite extension of Q; an algebraic number is an element of an algebraic number field. Algebraic number theory studies the arithmetic of algebraic number fields — the ring of integers in the number field, the ideals and units in the ring of integers, the extent to which unique factorization holds, and so on.

An abelian extension of a field is a Galois extension of the field with abelian Galois group. Class field theory describes the abelian extensions of a number field in terms of the arithmetic of the field.

INTRODUCTION

Algebraic number theory is a branch of number theory that uses the techniques of abstract algebra to study the integers, rational numbers, and their generalizations. Number-theoretic questions are expressed in terms of properties of algebraic objects such as algebraic number fields and their rings of integers, finite fields, and function fields. These properties, such as whether a ring admits unique factorization, the behavior of ideals, and the Galois groups of fields, can resolve questions of primary importance in number theory, like the existence of solutions to Diophantine equations.

HISTORY OF ALGEBRAIC NUMBER THEORY

Diophantus

The beginnings of algebraic number theory can be traced to Diophantine equations, named after the 3rd-century Alexandrian mathematician, Diophantus, who studied them and developed methods for the solution of some kinds of Diophantine equations. A typical Diophantine problem is to find two integers x and y such that their sum, and the sum of their squares, equal two given numbers A and B, respectively:

$$A = x + y$$
$$B = x^2 + y^2.$$

Diophantine equations have been studied for thousands of years. For example, the solutions to the quadratic Diophantine equation $x^2 + y^2 = z^2$ are given by the Pythagorean triples, originally solved by the Babylonians (c. 1800 BC). Solutions to linear Diophantine equations, such as $26x + 65y = 13$, may be found using the Euclidean algorithm (c. 5th century BC).

Diophantus's major work was the *Arithmetica*, of which only a portion has survived.

Fermat

Fermat's last theorem was first conjectured by Pierre de Fermat in 1637, famously in the margin of a copy of *Arithmetica* where he claimed he had a proof that was too large to fit in the margin. No successful proof was published until 1995 despite the efforts of countless mathematicians during the 358 intervening years. The unsolved problem stimulated the development of algebraic number theory in the 19th century and the proof of the modularity theorem in the 20th century.

Gauss

One of the founding works of algebraic number theory, the *Disquisitiones Arithmeticae* (Latin: *Arithmetical Investigations*) is a textbook of number theory written in Latin by Carl Friedrich Gauss in 1798 when Gauss was 21 and first published in 1801 when he was 24. In this book Gauss brings together results in number theory obtained by mathematicians such as Fermat, Euler, Lagrange and Legendre and adds important new results of his own. Before the *Disquisitiones* was published, number theory consisted of a collection of isolated theorems and conjectures. Gauss brought the work of his predecessors together with his own original work into a systematic framework, filled in gaps, corrected unsound proofs, and extended the subject in numerous ways.

The *Disquisitiones* was the starting point for the work of other nineteenth century European mathematicians including Ernst Kummer, Peter Gustav Lejeune Dirichlet and Richard Dedekind. Many of the annotations given by Gauss are in effect announcements of further research of his own, some of which remained unpublished. They must have appeared particularly cryptic to his contemporaries; we can now read them as containing the germs of the theories of L-functions and complex multiplication, in particular.

Dirichlet

In a couple of papers in 1838 and 1839 Peter Gustav Lejeune Dirichlet proved the first class number formula, for quadratic forms (later refined by his student Kronecker). The formula, which Jacobi called a result "touching the utmost of human acumen", opened the way for similar results regarding more general number fields. Based on his research of the structure of the unit group of quadratic fields, he proved the Dirichlet unit theorem, a fundamental result in algebraic number theory.

He first used the pigeonhole principle, a basic counting argument, in the proof of a theorem in diophantine approximation, later named after him Dirichlet's approximation theorem. He published important contributions to Fermat's last theorem, for which he proved the cases $n = 5$ and $n = 14$, and to the biquadratic reciprocity law. The Dirichlet divisor problem, for which he found the first results, is still an unsolved problem in number theory despite later contributions by other researchers.

Dedekind

Richard Dedekind's study of Lejeune Dirichlet's work was what led him to his later study of algebraic number fields and ideals. In 1863, he published Lejeune Dirichlet's lectures on number theory as *Vorlesungen über Zahlentheorie* ("Lectures on Number Theory") about which it has been written that:

"Although the book is assuredly based on Dirichlet's lectures, and although Dedekind himself referred to the book throughout his life as Dirichlet's, the book itself was entirely written by Dedekind, for the most part after Dirichlet's death." (Edwards 1983)

1879 and 1894 editions of the *Vorlesungen* included supplements introducing the notion of an ideal, fundamental to ring theory. (The word "Ring", introduced later by Hilbert, does not appear in Dedekind's work.) Dedekind defined an ideal as a subset of a set of numbers, composed of algebraic integers that satisfy polynomial equations with integer coefficients. The concept underwent further development in the hands of Hilbert and, especially, of Emmy Noether. Ideals generalize Ernst Eduard Kummer's ideal numbers, devised as part of Kummer's 1843 attempt to prove Fermat's Last Theorem.

Hilbert

David Hilbert unified the field of algebraic number theory with his 1897 treatise *Zahlbericht* (literally "report on numbers"). He also resolved a significant number-theory problem formulated by Waring in 1770. As with the finiteness theorem, he used an existence proof that shows there must be solutions for the problem rather than providing a mechanism to produce the answers. He then had little more to publish on the subject; but the emergence of Hilbert modular forms in the dissertation of a student means his name is further attached to a major area.

He made a series of conjectures on class field theory. The concepts were highly influential, and his own contribution lives on in the names of the Hilbert class field and of the Hilbert symbol of local class field theory. Results were mostly proved by 1930, after work by Teiji Takagi.

Artin

Emil Artin established the Artin reciprocity law in a series of papers (1924; 1927; 1930). This law is a general theorem in number theory that forms a central part of global class field theory. The term "reciprocity law" refers to a long line of more concrete number theoretic statements which it generalized, from the quadratic reciprocity law and the reciprocity laws of Eisenstein and Kummer to Hilbert's product formula for the norm symbol. Artin's result provided a partial solution to Hilbert's ninth problem.

Modern Theory

Around 1955, Japanese mathematicians Goro Shimura and Yutaka Taniyama observed a possible link between two apparently completely distinct, branches of mathematics, elliptic curves and modular forms. The resulting modularity theorem (at the time known as the Taniyama–Shimura conjecture) states that every elliptic curve is modular, meaning that it can be associated with a unique modular form.

It was initially dismissed as unlikely or highly speculative, and was taken more seriously when number theorist André Weil found evidence supporting it, but no proof; as a result the "astounding" conjecture was often known as the Taniyama–Shimura-Weil conjecture. It became a part of the Langlands programme, a list of important conjectures needing proof or disproof.

From 1993 to 1994, Andrew Wiles provided a proof of the modularity theorem for semistable elliptic curves, which, together with Ribet's theorem, provides a proof for Fermat's Last Theorem. Both Fermat's Last Theorem and the Modularity Theorem were almost universally considered inaccessible to proof by contemporaneous mathematicians (meaning, impossible or virtually impossible to prove using current knowledge). Wiles first announced his proof in June 1993 in a version that was soon recognized as having a serious gap in a key point. The proof was corrected by Wiles, in part via collaboration with Richard Taylor, and the final, widely accepted, version was released in September 1994, and formally published in 1995. The proof uses many techniques from algebraic geometry and number theory, and has many ramifications in these branches of mathematics. It also uses standard constructions of modern algebraic geometry, such as the category of schemes and Iwasawa theory, and other 20th-century techniques not available to Fermat.

BASIC NOTIONS

Failure of Unique Factorization

An important property of the ring of integers is that it satisfies the fundamental theorem of arithmetic, that every (positive) integer has a factorization into a product of prime numbers, and this factorization is unique up to the ordering of the factors. This may no longer be true in the ring of integers O of an algebraic number field K.

A *prime element* is an element p of O such that if p divides a product ab, then it divides one of the factors a or b. This property is closely related to primality in the integers, because any positive integer satisfying this property is either 1 or a prime number. However, it is strictly weaker. For example, −2 is not a prime number because it is negative, but it is a prime element. If factorizations into prime elements are permitted, then, even in the integers, there are alternative factorizations such as

$$6 = 2 \cdot 3 = (-2) \cdot (-3).$$

In general, if u is a unit, meaning a number with a multiplicative inverse in O, and if p is a prime element, then up is also a prime element. Numbers such as p and up are said to be *associate*. In the integers, the primes p and $-p$ are associate, but only one of these is positive. Requiring that prime numbers be positive selects a unique element from among a set of associated prime elements. When K is not the rational numbers, however, there is no analog of positivity. For example, in the Gaussian integers $\mathbf{Z}[i]$, the numbers $1 + 2i$ and $-2 + i$ are associate because the latter is the product of the former by i, but there is no way to single out one as being more canonical than the other. This leads to equations such as

$$5 = (1 + 2i)(1 - 2i) = (2 + i)(2 - i),$$

which prove that in $\mathbf{Z}[i]$, it is not true that factorizations are unique up to the order of the factors. For this reason, one adopts the definition of unique factorization used in unique factorization domains (UFDs). In a UFD, the prime elements occurring in a factorization are only expected to be unique up to units and their ordering.

However, even with this weaker definition, many rings of integers in algebraic number fields do not admit unique factorization. There is an algebraic obstruction called the ideal class group. When the ideal class group is trivial, the ring is a UFD. When it is not, there is a distinction between a prime element and an irreducible element. An *irreducible element* x is an element such that if $x = yz$, then either y or z is a unit. These are the elements that cannot be factored any further. Every element in O admits a factorization into irreducible elements, but it may admit more than one. This is because, while all prime elements are irreducible, some irreducible elements may not be prime. For example, consider the ring $\mathbf{Z}[\sqrt{-5}]$. In this ring, the numbers 3, $2 + \sqrt{-5}$ and $2 - \sqrt{-5}$ are irreducible. This means that the number 9 has two factorizations into irreducible elements,

$$9 = 3^2 = (2 + \sqrt{-5})(2 - \sqrt{-5}).$$

This equation shows that 3 divides the product $(2 + \sqrt{-5})(2 - \sqrt{-5}) = 9$. If 3 were a prime element, then it would divide $2 + \sqrt{-5}$ or $2 - \sqrt{-5}$, but it does not, because all elements divisible by 3 are of the form $3a + 3b\sqrt{-5}$. Similarly, $2 + \sqrt{-5}$ and $2 - \sqrt{-5}$ divide the product 3^2, but neither of these elements divides 3 itself, so neither of them are prime. As there is no sense in which the elements 3, $2 + \sqrt{-5}$ and $2 - \sqrt{-5}$ can be made equivalent, unique factorization fails in $\mathbf{Z}[\sqrt{-5}]$. Unlike the situation with units, where uniqueness could be repaired by weakening the definition, overcoming this failure requires a new perspective.

Factorization into Prime Ideals

If I is an ideal in O, then there is always a factorization

$$I = \mathfrak{p}_1^{e_1} \cdots \mathfrak{p}_t^{e_t},$$

where each $\mathfrak{p}_i$ is a prime ideal, and where this expression is unique up to the order of the factors. In particular, this is true if I is the principal ideal generated by a single element. This is the strongest sense in which the ring of integers of a general number field admits unique factorization. In the language of ring theory, it says that rings of integers are Dedekind domains.

When O is a UFD, every prime ideal is generated by a prime element. Otherwise, there are prime ideals which are not generated by prime elements. In $\mathbf{Z}[\sqrt{-5}]$, for instance, the ideal $(2, 1 + \sqrt{-5})$ is a prime ideal which cannot be generated by a single element.

Historically, the idea of factoring ideals into prime ideals was preceded by Ernst Kummer's introduction of ideal numbers. These are numbers lying in an extension field E of K. This extension field is now known as the Hilbert class field. By the principal ideal theorem, every prime ideal of O generates a principal ideal of the ring of integers of E. A generator of this principal ideal is called an ideal number. Kummer used these as a substitute for the failure of unique factorization in cyclotomic fields. These eventually led Richard Dedekind to introduce a forerunner of ideals and to prove unique factorization of ideals.

An ideal which is prime in the ring of integers in one number field may fail to be prime when extended to a larger number field. Consider, for example, the prime numbers. The corresponding ideals $p\mathbf{Z}$ are prime ideals of the ring $\mathbf{Z}$. However, when this ideal is extended to the Gaussian integers to get $p\mathbf{Z}[i]$, it may or may not be prime. For example, the factorization $2 = (1 + i)(1 - i)$ implies that

$$2\mathbf{Z}[i] = (1+i)\mathbf{Z}[i] \cdot (1-i)\mathbf{Z}[i] = ((1+i)\mathbf{Z}[i])^2;$$

note that because $1 + i = (1 - i) \cdot i$, the ideals generated by $1 + i$ and $1 - i$ are the same. A complete answer to the question of which ideals remain prime in the Gaussian integers is provided by Fermat's theorem on sums of two squares. It implies that for an odd prime number p, $p\mathbf{Z}[i]$ is a prime ideal if $p \equiv 3 \pmod 4$ and is not a prime ideal if $p \equiv 1 \pmod 4$. This, together with the observation that the ideal $(1 + i)\mathbf{Z}[i]$ is prime, provides a complete description of the prime ideals in the Gaussian integers. Generalizing this simple result to more general rings of integers is a basic problem in algebraic number theory. Class field theory accomplishes this goal when K is an abelian extension of $\mathbf{Q}$ (i.e. a Galois extension with abelian Galois group).

Ideal Class Group

Unique factorization fails if and only if there are prime ideals that fail to be principal. The object which measures the failure of prime ideals to be principal is called the ideal class group. Defining the ideal class group requires enlarging the set of ideals in a ring of algebraic integers so that they admit a group structure. This is done by generalizing ideals to fractional ideals. A fractional ideal is an additive subgroup J of K which is closed under multiplication by elements of O, meaning that $xJ \subseteq J$ if $x \in O$. All ideals of O are also fractional ideals. If I and J are fractional ideals, then the set IJ of all products of an element in I and an element in $J$is also a fractional ideal. This operation makes the set of non-zero fractional ideals into a group. The group identity is the ideal $(1) = O$, and the inverse of J is a (generalized) ideal quotient, $J^{-1} = (O : J) = \{ x \in K : xJ \subseteq O \}$.

The principal fractional ideals, meaning the ones of the form Ox where $x \in K^{\times}$, form a subgroup of the group of all non-zero fractional ideals. The quotient of the group of non-zero fractional ideals by this subgroup is the ideal class group. Two fractional ideals I and J represent the same element of the ideal class group if and only if there exists an element $x \in K$ such that $xI = J$. Therefore the ideal class group makes two fractional ideals equivalent if one is as close to being principal as the other is. The ideal class group is generally denoted Cl K, Cl O, or Pic O (with the last notation identifying it with the Picard group in algebraic geometry).

The number of elements in the class group is called the class number of K. The class number of Q(√-5) is 2. This means that there are only two ideal classes, the class of principal fractional ideals, and the class of a non-principal fractional ideal such as (2, 1 + √-5).

The ideal class group has another description in terms of divisors. These are formal objects which represent possible factorizations of numbers. The divisor group Div K is defined to be the free abelian group generated by the prime ideals of O. There is a group homomorphism from $K^{\times}$, the non-zero elements of K up to multiplication, to Div K. Suppose that $x \in K$ satisfies

$$(x) = \mathfrak{p}_1^{e_1} \cdots \mathfrak{p}_t^{e_t}.$$

Then div x is defined to be the divisor

$$\operatorname{div} x = \sum_{i=1}^{t} e_i [\mathfrak{p}_i].$$

The kernel of div is the group of units in O, while the cokernel is the ideal class group. In the language of homological algebra, this says that there is an exact sequence of abelian groups (written multiplicatively),

$$1 \to O^{\times} \to K^{\times} \xrightarrow{\text{div}} \operatorname{Div} K \to \operatorname{Cl} K \to 1.$$

Real and Complex Embeddings

Some number fields, such as **Q**(√2), can be specified as subfields of the real numbers. Others, such as **Q**(√−1), cannot. Abstractly, such a specification corresponds to a field homomorphism $K \to$ **R** or $K \to$ **C**. These are called real embeddings and complex embeddings, respectively. A real quadratic field **Q**(√d) is so-called because it admits two real embeddings and no complex embeddings. These are the field homomorphisms which send $\sqrt{d}$ to $\sqrt{d}$ and to $-\sqrt{d}$, respectively. Dually, an imaginary quadratic field $\mathbf{Q}(\sqrt{-d})$ admits no real embeddings and a conjugate pair of complex embeddings. One of these embeddings sends $\sqrt{-d}$ to $\sqrt{-d}$, while the other sends it to its complex conjugate.

Conventionally, the number of real embeddings of K is denoted r_1, while the number of conjugate pairs of complex embeddings is denoted r_2. The signature of K is the pair (r_1, r_2). It is a theorem that $r_1 + 2r_2 = d$, where d is the degree of K.

Considering all embeddings at once determines a function

$$M\colon K \to \mathbf{R}^{r_1} \oplus \mathbf{C}^{2r_2}.$$

This is called the Minkowski embedding. The subspace of the codomain fixed by complex conjugation is a real vector space of dimension d called Minkowski space. Because the Minkowski embedding is defined by field homomorphisms, multiplication of elements of K by an element $x \in K$ corresponds to multiplication by a diagonal matrix in the Minkowski embedding. The dot product on Minkowski space corresponds to the trace form $\langle x, y\rangle = \mathbf{Tr}(xy)$.

The image of O in Minkowski space is a d-dimensional lattice. If B is a basis for this lattice, then $\det B^{\mathrm{T}}B$ is the discriminant of O. The discriminant is denoted Δ or D. The covolume of the image of O is $\sqrt{|\Delta|}$.

Places

Real and complex embeddings can be put on the same footing as prime ideals by adopting a perspective based on valuations. Consider, for example, the integers. In addition to the usual absolute value function $|\cdot| : \mathbf{Q} \to \mathbf{R}$, there are p-adic absolute value functions $|\cdot|_p$: $\mathbf{Q} \to \mathbf{R}$, defined for each prime number p, which measure divisibility by p. Ostrowski's theorem states that these are all possible absolute value functions on $\mathbf{Q}$ (up to equivalence). Therefore absolute values are a common language to describe both the real embedding of $\mathbf{Q}$ and the prime numbers.

A **place** of an algebraic number field is an equivalence class of absolute value functions on K. There are two types of places. There is a $\mathfrak{p}$-adic absolute value for each prime ideal $\mathfrak{p}$ of O, and, like the p-adic absolute values, it measures divisibility. These are called finite places. The other type of place is specified using a real or complex embedding of K and the standard absolute value function on R or C. These are infinite places. Because absolute values are unable to distinguish between a complex embedding and its conjugate, a complex embedding and its conjugate determine the same place. Therefore there are r_1 real places and r_2 complex places. Because places encompass the primes, places are sometimes referred to as primes. When this is done, finite places are called finite primes and infinite places are called infinite primes. If v is a valuation corresponding to an absolute value, then one frequently writes $v \mid \infty$ to mean that v is an infinite place and $v \nmid \infty$ to mean that it is a finite place.

Considering all the places of the field together produces the adele ring of the number field. The adele ring allows one to simultaneously track all the data available using absolute values. This produces significant advantages in situations where the behavior at one place can affect the behavior at other places, as in the Artin reciprocity law.

Units

The integers have only two units, 1 and −1. Other rings of integers may admit more units. The Gaussian integers have four units, the previous two as well as $\pm i$. The Eisenstein integers $\mathbf{Z}[\exp(2\pi i / 3)]$ have six units. The integers in real quadratic number fields have infinitely many units. For example, in $\mathbf{Z}[\sqrt{3}]$, every power of $2 + \sqrt{3}$ is a unit, and all these powers are distinct.

In general, the group of units of O, denoted $O^\times$, is a finitely generated abelian group. The fundamental theorem of finitely generated abelian groups therefore implies that it is a direct sum of a torsion part and a free part. Reinterpreting this in the context of a number field, the torsion part consists of the roots of unity that lie in O. This group is cyclic. The free part is described by Dirichlet's unit theorem. This theorem says that rank of the free part is $r_1 + r_2 - 1$. Thus, for example, the only fields for which the rank of the free part is zero are $\mathbf{Q}$ and the imaginary quadratic fields. A more precise statement giving the structure of $O^\times \otimes_{\mathbf{Z}} \mathbf{Q}$ as a Galois module for the Galois group of $K/\mathbf{Q}$ is also possible.

The free part of the unit group can be studied using the infinite places of K. Consider the function

$$L: K^\times \to \mathbf{R}^{r_1+r_2}$$

defined by

$$L(x) = (\log |x|_v)_v,$$

where v varies over the infinite places of K and $|\cdot|_v$ is the absolute value associated with v. The function L is a homomorphism from $K^\times$ to a real vector space. It can be shown that the image of $O^\times$ is a lattice that spans the hyperplane defined by $x_1 + \cdots + x_{r_1+r_2} = 0$. The covolume of this lattice is the regulator of the number field. One of the simplifications made possible by working with the adele ring is that there is a single object, the idele class group, that describes both the quotient by this lattice and the ideal class group.

Zeta Function

The Dedekind zeta function of a number field, analogous to the Riemann zeta function is an analytic object which describes the behavior of prime ideals in K. When K is an abelian extension of $\mathbf{Q}$, Dedekind zeta functions are products of Dirichlet L-functions, with there being one factor for each Dirichlet character. The trivial character corresponds to the Riemann zeta function. When K is a Galois extension, the Dedekind zeta function is the Artin L-function of the regular representation of the Galois group of K, and it has a factorization in terms of irreducible Artin representations of the Galois group.

The zeta function is related to the other invariants described above by the class number formula.

Local Fields

Completing a number field K at a place w gives a complete field. If the valuation is archimedean, one gets **R** or **C**, if it is non-archimedean and lies over a prime p of the rationals, one gets a finite extension $K_w / \mathbf{Q}_p$: a complete, discrete valued field with finite residue field. This process simplifies the arithmetic of the field and allows the local study of problems. For example, the Kronecker–Weber theorem can be deduced easily from the analogous local statement. The philosophy behind the study of local fields is largely motivated by geometric methods. In algebraic geometry, it is common to study varieties locally at a point by localizing to a maximal ideal. Global information can then be recovered by gluing together local data. This spirit is adopted in algebraic number theory. Given a prime in the ring of algebraic integers in a number field, it is desirable to study the field locally at that prime. Therefore, one localizes the ring of algebraic integers to that prime and then completes the fraction field much in the spirit of geometry.

Major Results

Finiteness of the Class Group

One of the classical results in algebraic number theory is that the ideal class group of an algebraic number field K is finite. The order of the class group is called the class number, and is often denoted by the letter h.

Dirichlet's Unit Theorem

In mathematics, Dirichlet's unit theorem is a basic result in algebraic number theory due to Peter Gustav Lejeune Dirichlet. It determines the rank of the group of units in the ring O_K of algebraic integers of a number field K. The regulator is a positive real number that determines how "dense" the units are.

The statement is that the group of units is finitely generated and has rank (maximal number of multiplicatively independent elements) equal to

$$r = r_1 + r_2 - 1$$

where r_1 is the *number of real embeddings* and r_2 the *number of conjugate pairs of complex embeddings* of K. This characterisation of r_1 and r_2 is based on the idea that there will be as many ways to embed K in the complex number field as the degree $n = [K : \mathbf{Q}]$; these will either be into the real numbers, or pairs of embeddings related by complex conjugation, so that

$$n = r_1 + 2r_2.$$

Note that if K is Galois over **Q** then either r_1 is non-zero or r_2 is non-zero, but not both.

Other ways of determining r_1 and r_2 are

- use the primitive element theorem to write $K = \mathbf{Q}(\alpha)$, and then r_1 is the number of conjugates of α that are real, $2r_2$ the number that are complex;

- write the tensor product of fields $K \otimes_{\mathbf{Q}} \mathbf{R}$ as a product of fields, there being r_1 copies of **R** and r_2 copies of **C**.

As an example, if K is a quadratic field, the rank is 1 if it is a real quadratic field, and 0 if an imaginary quadratic field. The theory for real quadratic fields is essentially the theory of Pell's equation.

The rank is > 0 for all number fields besides **Q** and imaginary quadratic fields, which have rank 0. The 'size' of the units is measured in general by a determinant called the regulator. In principle a basis for the units can be effectively computed; in practice the calculations are quite involved when n is large.

The torsion in the group of units is the set of all roots of unity of K, which form a finite cyclic group. For a number field with at least one real embedding the torsion must therefore be only $\{1,-1\}$. There are number fields, for example most imaginary quadratic fields, having no real embeddings which also have $\{1,-1\}$ for the torsion of its unit group.

Totally real fields are special with respect to units. If L/K is a finite extension of number fields with degree greater than 1 and the units groups for the integers of L and K have the same rank then K is totally real and L is a totally complex quadratic extension. The converse holds too. (An example is K equal to the rationals and L equal to an imaginary quadratic field; both have unit rank 0.)

The theorem does not only applies to the maximal order O_K but to any order $O \subset O_K$.

There is a generalisation of the unit theorem by Helmut Hasse (and later Claude Chevalley) to describe the structure of the group of *S-units*, determining the rank of the unit group in localizations of rings of integers. Also, the Galois module structure of $\mathbf{Q} \oplus O_{K,S} \otimes_{\mathbf{Z}} \mathbf{Q}$ has been determined.

The Regulator

Suppose that $u_1,\dots,u_r$ are a set of generators for the unit group modulo roots of unity. If u is an algebraic number, write $u^1, \dots, u^{r+1}$ for the different embeddings into **R** or **C**, and set N_j to 1, resp. 2 if corresponding embedding is real, resp. complex. Then the r by $r + 1$ matrix whose entries are $N_j \log|u_i^j|$ has the property that the sum of any row is zero (because all units have norm 1, and the log of the norm is the sum of the entries of a row). This implies that the absolute value R of the determinant of the submatrix formed by deleting one column is independent of the column. The number R is called the **regulator** of the algebraic number field (it does not depend on the choice of generators u_i). It measures the "density" of the units: if the regulator is small, this means that there are "lots" of units.

The regulator has the following geometric interpretation. The map taking a unit u to the vector with entries $N_j \log|u^j|$ has image in the r-dimensional subspace of $\mathbf{R}^{r+1}$ consisting of all vector whose entries have sum 0, and by Dirichlet's unit theorem the image is a lattice in this subspace. The volume of a fundamental domain of this lattice is $R\sqrt{(r+1)}$.

The regulator of an algebraic number field of degree greater than 2 is usually quite cumbersome to calculate, though there are now computer algebra packages that can do it in many cases. It is usually much easier to calculate the product hR of the class number h and the regulator using the class number formula, and the main difficulty in calculating the class number of an algebraic number field is usually the calculation of the regulator.

Examples

- The regulator of an imaginary quadratic field, or of the rational integers, is 1 (as the determinant of a 0×0 matrix is 1).
- The regulator of a real quadratic field is the logarithm of its fundamental unit: for example, that of **Q**(√5) is log((√5 + 1)/2). This can be seen as follows. A fundamental unit is (√5 + 1)/2, and its images under the two embeddings into **R** are (√5 + 1)/2 and (−√5 + 1)/2. So the r by $r + 1$ matrix is

$$\left[1 \times \log\left|\frac{\sqrt{5}+1}{2}\right|, \quad 1 \times \log\left|\frac{-\sqrt{5}+1}{2}\right|\right].$$

- The regulator of the cyclic cubic field **Q(α),** where **α** is a root of $x^3 + x^2 - 2x - 1$, is approximately 0.5255. A basis of the group of units modulo roots of unity is $\{\varepsilon_1, \varepsilon_2\}$ where $\varepsilon_1 = \alpha^2 + \alpha - 1$ and $\varepsilon_2 = 2 - \alpha^2$.

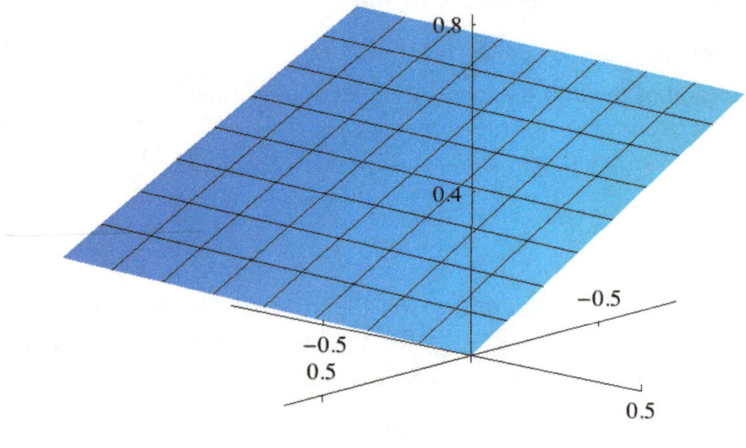

A fundamental domain in logarithmic space of the group of units of the cyclic cubic field K obtained by adjoining to **Q** a root of $f(x) = x^3 + x^2 - 2x - 1$. If **α** denotes a root of $f(x)$, then a set of fundamental units is $\{\varepsilon_1, \varepsilon_2\}$ where $\varepsilon_1 = \alpha^2 + \alpha - 1$ and $\varepsilon_2 = 2 - \alpha^2$. The area of the fundamental domain is approximately 0.910114, so the regulator of K is approximately 0.525455.

Higher Regulators

A 'higher' regulator refers to a construction for a function on an algebraic K-group with index $n > 1$ that plays the same role as the classical regulator does for the group of units, which is a group K_1. A theory of such regulators has been in development, with work of Armand Borel and others. Such higher regulators play a role, for example, in the Beilinson conjectures, and are expected to occur in evaluations of certain L-functions at integer values of the argument.

Stark Regulator

The formulation of Stark's conjectures led Harold Stark to define what is now called the Stark regulator, similar to the classical regulator as a determinant of logarithms of units, attached to any Artin representation.

p-adic Regulator

Let K be a number field and for each prime P of K above some fixed rational prime p, let U_P denote the local units at P and let $U_{1,P}$ denote the subgroup of principal units in U_P. Set

$$U_1 = \prod_{P|p} U_{1,P}.$$

Then let E_1 denote the set of global units ε that map to U_1 via the diagonal embedding of the global units in E.

Since E_1 is a finite-index subgroup of the global units, it is an abelian group of rank $r_1 + r_2 - 1$. The ***p*-adic regulator** is the determinant of the matrix formed by the p-adic logarithms of the generators of this group. *Leopoldt's conjecture* states that this determinant is non-zero.

Reciprocity Laws

In terms of the Legendre symbol, the law of quadratic reciprocity for positive odd primes states

$$\left(\frac{p}{q}\right)\left(\frac{q}{p}\right) = (-1)^{\frac{p-1}{2}\frac{q-1}{2}}.$$

A reciprocity law is a generalization of the law of quadratic reciprocity.

There are several different ways to express reciprocity laws. The early reciprocity laws found in the 19th century were usually expressed in terms of a power residue symbol (p/q) generalizing the quadratic reciprocity symbol, that describes when a prime number is an nth power residue modulo another prime, and gave a relation between (p/q) and (q/p). Hilbert reformulated the reciprocity laws as saying that a product over p of Hilbert symbols $(a,b/p)$, taking values in roots of unity, is equal to 1. Artin's reformulated reciprocity law states that the Artin symbol from ideals (or ideles) to elements of a Galois group is trivial on a certain subgroup. Several more recent generalizations express reciprocity laws using cohomology of groups or representations of adelic groups or algebraic K-groups, and their relationship with the original quadratic reciprocity law can be hard to see.

Class number Formula

In number theory, the class number formula relates many important invariants of a number field to a special value of its Dedekind zeta function.

General Statement of the Class Number Formula

We start with the following data:

- K is a number field.
- $[K : \mathbf{Q}] = n = r_1 + 2r_2$, where r_1 denotes the number of real embeddings of K, and $2r_2$ is the number of complex embeddings of K.
- ${}_K(s)$ is the Dedekind zeta function of K.
- h_K is the class number, the number of elements in the ideal class group of K.
- Reg_K is the regulator of K.
- w_K is the number of roots of unity contained in K.
- D_K is the discriminant of the extension $K/\mathbf{Q}$.

Then:

Theorem (Class Number Formula). $\zeta_K(s)$ converges absolutely for $\text{Re}(s) > 1$ and extends to a meromorphic function defined for all complex s with only one simple pole at $s = 1$, with residue

$$\lim_{s\to 1}(s-1)\zeta_K(s) = \frac{2^{r_1}\cdot(2\pi)^{r_2}\cdot h_K\cdot \text{Reg}_K}{w_K\cdot\sqrt{|D_K|}}$$

This is the most general "class number formula". In particular cases, for example when K is a cyclotomic extension of $\mathbf{Q}$, there are particular and more refined class number formulas.

Proof

The idea of the proof of the class number formula is most easily seen when $K = \mathbf{Q}(i)$. In this case, the ring of integers in K is the Gaussian integers.

An elementary manipulation shows that the residue of the Dedekind zeta function at $s = 1$ is the average of the coefficients of the Dirichlet series representation of the Dedekind zeta function. The n-th coefficient of the Dirichlet series is essentially the number of representations of n as a sum of two squares of nonnegative integers. So one can compute the residue of the Dedekind zeta function at $s = 1$ by computing the average number of representations. As in the article on the Gauss circle problem, one can compute this by approximating the number of lattice points inside of a quarter circle centered at the origin, concluding that the residue is one quarter of pi.

The proof when K is an arbitrary imaginary quadratic number field is very similar.

In the general case, by Dirichlet's unit theorem, the group of units in the ring of integers of K is infinite. One can nevertheless reduce the computation of the residue to a lattice point counting problem using the classical theory of real and complex embeddings and approximate the number of lattice points in a region by the volume of the region, to complete the proof.

Dirichlet Class Number Formula

Peter Gustav Lejeune Dirichlet published a proof of the class number formula for quadratic fields in 1839, but it was stated in the language of quadratic forms rather than classes of ideals. It appears that Gauss already knew this formula in 1801.

This exposition follows Davenport.

Let d be a fundamental discriminant, and write $h(d)$ for the number of equivalence classes of quadratic forms with discriminant d. Let $\chi = \left(\frac{d}{m}\right)$ be the Kronecker symbol. Then χ is a Dirichlet character. Write $L(s,\chi)$ for the Dirichlet L-series based on χ. For $d > 0$, let $t > 0$, $u > 0$ be the solution to the Pell equation $t^2 - du^2 = 4$ for which u is smallest, and write

$$\epsilon = \frac{1}{2}(t + u\sqrt{d}).$$

(Then ε is either a fundamental unit of the real quadratic field $\mathbb{Q}(\sqrt{d})$ or the square of a fundamental unit.) For $d < 0$, write w for the number of automorphs of quadratic forms of discriminant d; that is,

$$w = \begin{cases} 2, & d < -4; \\ 4, & d = -4; \\ 6, & d = -3. \end{cases}$$

Then Dirichlet showed that

$$h(d) = \begin{cases} \dfrac{w\sqrt{|d|}}{2\pi} L(1,\chi), & d < 0; \\ \dfrac{\sqrt{d}}{\ln \epsilon} L(1,\chi), & d > 0. \end{cases}$$

This is a special case of Theorem 1 above: for a quadratic field K, the Dedekind zeta function is just $\zeta_K(s) = \zeta(s)L(s,\chi)$, and the residue is $L(1,\chi)$. Dirichlet also showed that the L-series can be written in a finite form, which gives a finite form for the class number. Suppose χ is primitive with prime conductor q. Then

$$L(1,\chi) = \begin{cases} -\dfrac{\pi}{q^{3/2}} \sum_{m=1}^{q-1} m \left(\dfrac{m}{q}\right), & q \equiv 3 \mod 4; \\ -\dfrac{1}{q^{1/2}} \sum_{m=1}^{q-1} \left(\dfrac{m}{q}\right) \ln 2 \sin \dfrac{m\pi}{q}, & q \equiv 1 \mod 4. \end{cases}$$

Galois Extensions of the Rationals

If K is a Galois extension of **Q**, the theory of Artin L-functions applies to $\zeta_K(s)$. It has one factor of the Riemann zeta function, which has a pole of residue one, and the quotient is regular at $s = 1$. This means that the right-hand side of the class number formula can be equated to a left-hand side

$L(1,\rho)^{\dim}$

with running over the classes of irreducible non-trivial complex linear representations of Gal(K/**Q**) of dimension dim(ρ). That is according to the standard decomposition of the regular representation.

Abelian Extensions of the Rationals

This is the case of the above, with Gal(K/**Q**) an abelian group, in which all the can be replaced by Dirichlet characters (via class field theory) for some modulus f called the conductor. Therefore all the $L(1)$ values occur for Dirichlet L-functions, for which there is a classical formula, involving logarithms.

By the Kronecker–Weber theorem, all the values required for an analytic class number formula occur already when the cyclotomic fields are considered. In that case there is a further formulation possible, as shown by Kummer. The regulator, a calculation of volume in 'logarithmic space' as divided by the logarithms of the units of the cyclotomic field, can be set against the quantities from the $L(1)$ recognisable as logarithms of cyclotomic units. There result formulae stating that the class number is determined by the index of the cyclotomic units in the whole group of units.

In Iwasawa theory, these ideas are further combined with Stickelberger's theorem.

Related Areas

Algebraic number theory interacts with many other mathematical disciplines. It uses tools from homological algebra. Via the analogy of function fields vs. number fields, it relies on techniques and ideas from algebraic geometry. Moreover, the study of higher-dimensional schemes over **Z** instead of number rings is referred to as arithmetic geometry. Algebraic number theory is also used in the study of arithmetic hyperbolic 3-manifolds.

ADELIC ALGEBRAIC GROUP

In abstract algebra, an adelic algebraic group is a semitopological group defined by an algebraic group G over a number field K, and the adele ring $A = A(K)$ of K. It consists of the points of G having values in A; the definition of the appropriate topology is straightforward only in case G is a linear algebraic group. In the case of G an abelian variety it presents a technical obstacle, though it is known that the concept is potentially useful in connection with Tamagawa numbers. Adelic algebraic groups are widely used in number theory, particularly for the theory of automorphic representations, and the arithmetic of quadratic forms.

In case G is a linear algebraic group, it is an affine algebraic variety in affine N-space. The topology on the adelic algebraic group G(A) is taken to be the subspace topology in A^N, the Cartesian product of N copies of the adele ring.

Ideles

An important example, the idele group *I*(*K*), is the case of $G = GL_1$. Here the set of ideles (also *idèles* /dlz/) consists of the invertible adeles; but the topology on the idele group is *not* their topology as a subset of the adeles. Instead, considering that GL_1 lies in two-dimensional affine space as the 'hyperbola' defined parametrically by

$\{(t, t^{-1})\}$,

the topology correctly assigned to the idele group is that induced by inclusion in A^2; composing with a projection, it follows that the ideles carry a finer topology than the subspace topology from A.

Inside A^N, the product K^N lies as a discrete subgroup. This means that $G(K)$ is a discrete subgroup of $G(A)$, also. In the case of the idele group, the quotient group

$I(K)/K^{\times}$

is the idele class group. It is closely related to (though larger than) the ideal class group. The idele class group is not itself compact; the ideles must first be replaced by the ideles of norm 1, and then the image of those in the idele class group is a compact group; the proof of this is essentially equivalent to the finiteness of the class number.

The study of the Galois cohomology of idele class groups is a central matter in class field theory. Characters of the idele class group, now usually called Hecke characters, give rise to the most basic class of L-functions.

Tamagawa Numbers

For more general *G*, the Tamagawa number is defined (or indirectly computed) as the measure of

$G(A)/G(K)$.

Tsuneo Tamagawa's observation was that, starting from an invariant differential form ω on *G*, defined *over K*, the measure involved was well-defined: while ω could be replaced by $c\omega$ with c a non-zero element of *K*, the product formula for valuations in *K* is reflected by the independence from c of the measure of the quotient, for the product measure constructed from ω on each effective factor. The computation of Tamagawa numbers for semisimple groups contains important parts of classical quadratic form theory.

History of the Terminology

Historically the *idèles* were introduced by Chevalley (1936) under the name "élément idéal", which is "ideal element" in French, which Chevalley (1940) then abbreviated to "idèle" following a suggestion of Hasse. (In these papers he also gave the ideles a non-Hausdorff topology.) This was to formulate class field theory for infinite extensions in terms of topological groups. Weil (1938) defined (but did not name) the ring of adeles in the function field case and pointed out that Chevalley's group of *Idealelemente* was the group of invertible elements of this ring. Tate (1950) defined the ring of adeles as a restricted direct product, though he called its elements "valuation vectors" rather than adeles. Chevalley (1951) defined the ring of adeles in the function field case, under the name "repartitions". The term *adèle* (short for additive idèles, and also a French woman's name) was in use shortly afterwards (Jaffard 1953) and may have been introduced by André Weil. The general construction of adelic algebraic groups by Ono (1957) followed the algebraic group theory founded by Armand Borel and Harish-Chandra.

CLASS FIELD THEORY

In mathematics, class field theory is a major branch of algebraic number theory that studies abelian extensions of number fields and function fields of curves over finite fields and arithmetic properties of such abelian extensions. A general name for such fields is global fields, or one-dimensional global fields.

The theory takes its name from the fact that it provides a one-to-one correspondence between finite abelian extensions of a fixed global field and appropriate classes of ideals of the field or open subgroups of the idele class group of the field. For example, the Hilbert class field, which is the maximal unramified abelian extension of a number field, corresponds to a very special class of ideals. Class field theory also includes a reciprocity homomorphism, which acts from the idele class group of a global field, i.e. the quotient of the ideles by the multiplicative group of the field, to the Galois group of the maximal abelian extension of the global field. Each open subgroup of the idele class group of a global field is the image with respect to the norm map from the corresponding class field extension down to the global field.

A standard method since the 1930s is to develop local class field theory, which describes abelian extensions of completions of a local field, and then use it to construct global class field theory.

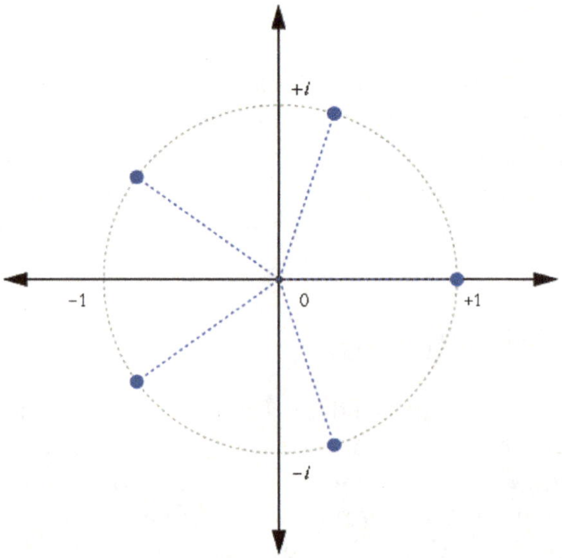

The fifth roots of unity in the complex plane. Adjoining these roots to the rational numbers generates an abelian extension.

Formulation in Contemporary Language

In modern language there is a *maximal* abelian extension A of K, which will be of infinite degree over K; and associated to A a Galois group G, which will be a pro-finite group, so a compact topological group, and also abelian. The central aim of the theory is to describe G in terms of K. In particular to establish a one-to-one correspondence between finite abelian extensions of K and their norm groups in an appropriate object for K, such as the multiplicative group in the case of local fields with finite residue field and the idele class group in the case of global fields, as well as to describe those norm groups directly, e.g., such as open subgroups of finite index. The finite abelian extension corresponding to such a subgroup is called a class field, which gave the name to the theory.

The fundamental result of class field theory states that the group G is naturally isomorphic to the profinite completion of the idele class group C_K of K with respect to the natural topology on C_K related to the specific structure of the field K. Equivalently, for any finite Galois extension L of K, there is an isomorphism

$$\mathrm{Gal}(L / K)^{ab} \to C_K / N_{L/K}\, C_L$$

of the maximal abelian quotient of the Galois group of the extension with the quotient of the idele class group of K by the image of the norm of the idele class group of L.

For some small fields, such as the field of rational numbers Q or its quadratic imaginary extensions there is a more detailed theory which provides more information. For example, the abelianized absolute Galois group G of Q is (naturally isomorphic to) an infinite product of the group of units of the p-adic integers taken over all prime numbers p, and the corresponding maximal abelian extension of the rationals is the field generated by all roots of unity. This is known as the Kronecker–Weber theorem, originally conjectured by Leopold Kronecker. In this case the reciprocity isomorphism of class field theory (or Artin reciprocity map) also admits an explicit description due to the Kronecker–Weber theorem. Let us denote with

$$\mu_\infty \subset \mathbb{C}^\times$$

the group of all roots of unity, i.e. the torsion subgroup. The Artin reciprocity map is given by

$$\hat{\mathbb{Z}}^\times \to G_{\mathbb{Q}}^{ab} = \mathrm{Gal}(\mathbb{Q}(\mu_\infty)/\mathbb{Q}), \quad x \mapsto (\zeta \mapsto \zeta^x),$$

when it is arithmetically normalized, or given by

$$\hat{\mathbb{Z}}^\times \to G_{\mathbb{Q}}^{ab} = \mathrm{Gal}(\mathbb{Q}(\mu_\infty)/\mathbb{Q}), \quad x \mapsto (\zeta \mapsto \zeta^{-x}),$$

if it is geometrically normalized. However, principal constructions of such more detailed theories for small algebraic number fields are not extendable to the general case of algebraic number fields, and different conceptual principles are in use in the general class field theory.

The standard method to construct the reciprocity homomorphism is to first construct the local reciprocity isomorphism from the multiplicative group of the completion of a global field to the Galois group of its maximal abelian extension (this is done inside local class field theory) and then prove that the product of all such local reciprocity maps when defined on the idele group of the global field is trivial on the image of the multiplicative group of the global field. The latter property is called the *global reciprocity law* and is a far reaching generalization of the Gauss quadratic reciprocity law.

One of the methods to construct the reciprocity homomorphism uses class formation.

There are methods which use cohomology groups, in particular the Brauer group, and there are methods which do not use cohomology groups and are very explicit and good for applications.

Prime Ideals

More than just the abstract description of *G*, it is essential for the purposes of number theory to understand how prime ideals decompose in the abelian extensions. The description is in terms of Frobenius elements, and generalises in a far-reaching way the quadratic reciprocity law that gives full information on the decomposition of prime numbers in quadratic fields. The class field theory project included the 'higher reciprocity laws' (cubic reciprocity) and so on.

Applications

Class field theory is used to prove Artin-Verdier duality. It is also used in Iwasawa theory.

Generalizations of Class Field Theory

One natural development in number theory is to understand and construct nonabelian class field theories which provide information about general Galois extensions of global fields. Often, the Langlands correspondence is viewed as a nonabelian class field theory and indeed when fully established it will contain a very rich theory of nonabelian Galois extensions of global fields. However, the Langlands correspondence does not include as much arithmetical information about finite Galois extensions as class field theory does in the abelian case. It also does not include an analog of the existence theorem in class field theory, i.e. the concept of class fields is absent in the Langlands correspondence. There are several other nonabelian theories, local and global, which provide alternative to the Langlands correspondence point of view.

Another natural development in arithmetic geometry is to understand and construct class field theory which describes abelian extensions of higher local and global fields. The latter come as function fields of schemes of finite type over integers and their appropriate localization and completions. The theory is referred to as *higher local and global class field theory* or *geometric class field theory*. It uses algebraic K-theory and appropriate Milnor K-groups replace K_1 which is in use in one-dimensional class field theory. Higher local and global class field theory was developed by A. Parshin, Kazuya Kato, Ivan Fesenko, Spencer Bloch, Shuji Saito and other mathematicians. There are attempts to develop higher global class field theory without using algebraic K-theory

(G. Wiesend), but his approach does not involve higher local class field theory and a compatibility between the local and global theories.

History

The origins of class field theory lie in the quadratic reciprocity law proved by Gauss. The generalisation took place as a long-term historical project, involving quadratic forms and their 'genus theory', work of Ernst Kummer and Leopold Kronecker/Kurt Hensel on ideals and completions, the theory of cyclotomic and Kummer extensions.

The first two class field theories were very explicit cyclotomic and complex multiplication class field theories. They used additional structures: in the case of the field of rational numbers they use roots of unity, in the case of imaginary quadratic extensions of the field of rational numbers they use elliptic curves with complex multiplication and their points of finite order. Much later, the theory of Shimura provided another very explicit class field theory for a class of algebraic number fields. All these very explicit theories cannot be extended to work over arbitrary number field. In positive characteristic p Kawada and Satake used Witt duality to get a very easy description of the p-part of the reciprocity homomorphism.

However, general class field theory used different concepts and its constructions work over every global field.

The famous problems of David Hilbert stimulated further development, which led to the reciprocity laws, and proofs by Teiji Takagi, Phillip Furtwängler, Emil Artin, Helmut Hasse and many others. The crucial Takagi existence theorem was known by 1920 and all the main results by about 1930. One of the last classical conjectures to be proved was the principalisation property. The first proofs of class field theory used substantial analytic methods. In the 1930s and subsequently the use of infinite extensions and the theory of Wolfgang Krull of their Galois groups was found increasingly useful. It combines with Pontryagin duality to give a clearer if more abstract formulation of the central result, the Artin reciprocity law. An important step was the introduction of ideles by Claude Chevalley in 1930s. Their use replaced the classes of ideals and essentially clarified and simplified structures that describe abelian extensions of global fields. Most of the central results were proved by 1940.

Later the results were reformulated in terms of group cohomology, which became a standard way to learn class field theory for several generations of number theorists. One drawback of the cohomological method is its relative inexplicitness. As the result of local contributions by Bernard Dwork, John Tate, Michiel Hazewinkel and a local and global reinterpretation by Jürgen Neukirch and also in relation to the work on explicit reciprocity formulas by many mathematicians, a very explicit and cohomology free presentation of class field theory was established in the nineties, see e.g. the book of Neukirch.

KUMMER THEORY

In abstract algebra and number theory, Kummer theory provides a description of certain types of field extensions involving the adjunction of nth roots of elements of the base field. The theory was originally developed by Ernst Eduard Kummer around the 1840s in his pioneering work on Fermat's last theorem. The main statements do not depend on the nature of the field - apart from its characteristic, which should not divide the integer n – and therefore belong to abstract algebra. The theory of cyclic extensions of the field K when the characteristic of K does divide n is called Artin–Schreier theory.

Kummer theory is basic, for example, in class field theory and in general in understanding abelian extensions; it says that in the presence of enough roots of unity, cyclic extensions can be understood in terms of extracting roots. The main burden in class field theory is to dispense with extra roots of unity ('descending' back to smaller fields); which is something much more serious.

Kummer Extensions

A Kummer extension is a field extension L/K, where for some given integer $n > 1$ we have

- K contains n distinct nth roots of unity (i.e., roots of X^n-1)
- L/K has abelian Galois group of exponent n.

For example, when $n = 2$, the first condition is always true if K has characteristic $\neq$ 2. The Kummer extensions in this case include quadratic extensions $L = K(\sqrt{a})$ where a in K is a non-square element. By the usual solution of quadratic equations, any extension of degree 2 of K has this form. The Kummer extensions in this case also include biquadratic extensions and more general multiquadratic extensions. When K has characteristic 2, there are no such Kummer extensions.

Taking $n = 3$, there are no degree 3 Kummer extensions of the rational number field $\mathbf{Q}$, since for three cube roots of 1 complex numbers are required. If one takes L to be the splitting field of $X^3 - a$ over $\mathbf{Q}$, where a is not a cube in the rational numbers, then L contains a subfield K with three cube roots of 1; that is because if α and β are roots of the cubic polynomial, we shall have $(\alpha/\beta)^3 = 1$ and the cubic is a separable polynomial. Then L/K is a Kummer extension.

More generally, it is true that when K contains n distinct nth roots of unity, which implies that the characteristic of K doesn't divide n, then adjoining to K the nth root of any element a of K creates a Kummer extension (of degree m, for some m dividing n). As the splitting field of the polynomial $X^n - a$, the Kummer extension is necessarily Galois, with Galois group that is cyclic of order m. It is easy to track the Galois action via the root of unity in front of $\sqrt[n]{a}$.

Kummer theory provides converse statements. When K contains n distinct nth roots of unity, it states that any abelian extension of K of exponent dividing n is formed by extraction of roots of elements of K. Further, if $K^\times$ denotes the multiplicative group of non-zero elements of K, abelian extensions of K of exponent n correspond bijectively with subgroups of

$$K^\times/(K^\times)^n,$$

that is, elements of $K^\times$ modulo nth powers. The correspondence can be described explicitly as follows. Given a subgroup

$$\Delta \subseteq K^{\times}/(K^{\times})^n,$$

the corresponding extension is given by

$$K(\Delta^{1/n}),$$

where $\Delta^{1/n} = \{\sqrt[n]{a} : a \in K^{\times}, a \cdot (K^{\times})^n \in \Delta\}$. In fact it suffices to adjoin nth root of one representative of each element of any set of generators of the group Δ. Conversely, if L is a Kummer extension of K, then Δ is recovered by the rule

$$\Delta = (K^{\times} \cap (L^{\times})^n)/(K^{\times})^n.$$

In this case there is an isomorphism

$$\Delta \cong \mathrm{Hom_c}(\mathrm{Gal}(L/K), \mu_n)$$

given by

$$a \mapsto \left(\sigma \mapsto \frac{\sigma(\alpha)}{\alpha}\right),$$

where α is any nth root of a in L. Here μ_n denotes the multiplicative group of nth roots of unity (which belong to K) and $\mathrm{Hom_c}(\mathrm{Gal}(L/K), \mu_n)$ is the group of continuous homomorphisms from $\mathrm{Gal}(L/K)$ equipped with Krull topology to μ_n with discrete topology (with group operation given by pointwise multiplication). This group (with discrete topology) can also be viewed as Pontryagin dual of $\mathrm{Gal}(L/K)$, assuming we regard μ_n as a subgroup of circle group. If the extension L/K is finite, then $\mathrm{Gal}(L/K)$ is a finite discrete group and we have

$$\Delta \cong \mathrm{Hom}(\mathrm{Gal}(L/K), \mu_n) \cong \mathrm{Gal}(L/K),$$

however the last isomorphism isn't natural.

Generalizations

Suppose that G is a profinite group acting on a module A with a surjective homomorphism π from the G-module A to itself. Suppose also that G acts trivially on the kernel C of π and that the first cohomology group $H^1(G,A)$ is trivial. Then the exact sequence of group cohomology shows that there is an isomorphism between $A^G/\pi(A^G)$ and $\mathrm{Hom}(G,C)$.

Kummer theory is the special case of this when A is the multiplicative group of the separable closure of a field k, G is the Galois group, π is the nth power map, and C the group of nth roots of unity. Artin–Schreier theory is the special case when A is the additive group of the separable closure of a field k of positive characteristic p, G is the Galois group, π is the Frobenius map, and C the finite field of order p. Taking A to be a ring of truncated Witt vectors gives Witt's generalization of Artin–Schreier theory to extensions of exponent dividing p^n.

IDEAL CLASS GROUP

In mathematics, for a field K an ideal class group (or class group) is the quotient group J_K/P_K where J_K is the group of fractional ideals of K and P_K is its subgroup of principal ideals of K. The extent to which unique factorization fails in the ring of integers of an algebraic number field (or more generally any Dedekind domain) can be described by the ideal class group (or class group). If this group is finite (as it is in the case of the ring of integers of a number field), then the order of the group is called the class number. The multiplicative theory of a Dedekind domain is intimately tied to the structure of its class group. For example, the class group of a Dedekind domain is trivial if and only if the ring is a unique factorization domain.

History and Origin of the Ideal Class Group

Ideal class groups (or, rather, what were effectively ideal class groups) were studied some time before the idea of an ideal was formulated. These groups appeared in the theory of quadratic forms: in the case of binary integral quadratic forms, as put into something like a final form by Gauss, a composition law was defined on certain equivalence classes of forms. This gave a finite abelian group, as was recognized at the time.

Later Kummer was working towards a theory of cyclotomic fields. It had been realized (probably by several people) that failure to complete proofs in the general case of Fermat's last theorem by factorization using the roots of unity was for a very good reason: a failure of the fundamental theorem of arithmetic to hold in the rings generated by those roots of unity was a major obstacle. Out of Kummer's work for the first time came a study of the obstruction to the factorization. We now recognize this as part of the ideal class group: in fact Kummer had isolated the p-torsion in that group for the field of p-roots of unity, for any prime number p, as the reason for the failure of the standard method of attack on the Fermat problem.

Somewhat later again Dedekind formulated the concept of ideal, Kummer having worked in a different way. At this point the existing examples could be unified. It was shown that while rings of algebraic integers do not always have unique factorization into primes (because they need not be principal ideal domains), they do have the property that every proper ideal admits a unique factorization as a product of prime ideals (that is, every ring of algebraic integers is a Dedekind domain). The size of the ideal class group can be considered as a measure for the deviation of a ring from being a principal ideal domain; a ring is a principal domain if and only if it has a trivial ideal class group.

Definition

If R is an integral domain, define a relation $\sim$ on nonzero fractional ideals of R by $I \sim J$ whenever there exist nonzero elements a and b of R such that $(a)I = (b)J$. (Here the notation (a) means the principal ideal of R consisting of all the multiples of a.) It is easily shown that this is an equivalence relation. The equivalence classes are called the *ideal classes* of R. Ideal classes can be multiplied: if $[I]$ denotes the equivalence class of the ideal I, then the multiplication $[I][J] = [IJ]$ is well-defined and commutative. The principal ideals form the ideal class [R] which serves as an identity element for this multiplication. Thus a class [I] has an inverse [J] if and only if there is an ideal J such that IJ is a principal ideal. In general, such a J may not exist and consequently the set of ideal classes of R may only be a monoid.

However, if R is the ring of algebraic integers in an algebraic number field, or more generally a Dedekind domain, the multiplication defined above turns the set of fractional ideal classes into an abelian group, the ideal class group of R. The group property of existence of inverse elements follows easily from the fact that, in a Dedekind domain, every non-zero ideal (except R) is a product of prime ideals.

Properties

The ideal class group is trivial (i.e. has only one element) if and only if all ideals of R are principal. In this sense, the ideal class group measures how far R is from being a principal ideal domain, and hence from satisfying unique prime factorization (Dedekind domains are unique factorization domains if and only if they are principal ideal domains).

The number of ideal classes (the **class number** of R) may be infinite in general. In fact, every abelian group is isomorphic to the ideal class group of some Dedekind domain. But if R is in fact a ring of algebraic integers, then the class number is always *finite*. This is one of the main results of classical algebraic number theory.

Computation of the class group is hard, in general; it can be done by hand for the ring of integers in an algebraic number field of small discriminant, using Minkowski's bound. This result gives a bound, depending on the ring, such that every ideal class contains an ideal norm less than the bound. In general the bound is not sharp enough to make the calculation practical for fields with large discriminant, but computers are well suited to the task.

The mapping from rings of integers R to their corresponding class groups is functorial, and the class group can be subsumed under the heading of algebraic K-theory, with $K_0(R)$ being the functor assigning to R its ideal class group; more precisely, $K_0(R) = \mathbf{Z} \times C(R)$, where $C(R)$ is the class group. Higher K groups can also be employed and interpreted arithmetically in connection to rings of integers.

Relation with the Group of Units

It was remarked above that the ideal class group provides part of the answer to the question of how much ideals in a Dedekind domain behave like elements. The other part of the answer is provided by the multiplicative group of units of the Dedekind domain, since passage from principal ideals to their generators requires the use of units (and this is the rest of the reason for introducing the concept of fractional ideal, as well):

Define a map from $K^{\times}$ to the set of all nonzero fractional ideals of R by sending every element to the principal (fractional) ideal it generates. This is a group homomorphism; its kernel is the group of units of R, and its cokernel is the ideal class group of R. The failure of these groups to be trivial is a measure of the failure of the map to be an isomorphism: that is the failure of ideals to act like ring elements, that is to say, like numbers.

Examples of Ideal Class Groups

- The rings **Z**, **Z**[ω], and **Z**[*i*], where ω is a cube root of 1 and *i* is a fourth root of 1 (i.e. a square root of −1), are all principal ideal domains (and in fact are all Euclidean domains), and so have class number 1: that is, they have trivial ideal class groups.
- If *k* is a field, then the polynomial ring $k[X_1, X_2, X_3, ...]$ is an integral domain. It has a countably infinite set of ideal classes.

Class Numbers of Quadratic Fields

If *d* is a square-free integer (a product of distinct primes) other than 1, then $\mathbf{Q}(\sqrt{d})$ is a quadratic extension of **Q**. If $d < 0$, then the class number of the ring *R* of algebraic integers of $\mathbf{Q}(\sqrt{d})$ is equal to 1 for precisely the following values of *d*: $d = -1, -2, -3, -7, -11, -19, -43, -67$, and -163. This result was first conjectured by Gauss and proven by Kurt Heegner, although Heegner's proof was not believed until Harold Stark gave a later proof in 1967. This is a special case of the famous class number problem.

If, on the other hand, $d > 0$, then it is unknown whether there are infinitely many fields $\mathbf{Q}(\sqrt{d})$ with class number 1. Computational results indicate that there are a great many such fields. However, it is not even known if there are infinitely many number fields with class number 1.

For $d < 0$, the ideal class group of $\mathbf{Q}(\sqrt{d})$ is isomorphic to the class group of integral binary quadratic forms of discriminant equal to the discriminant of $\mathbf{Q}(\sqrt{d})$. For $d > 0$, the ideal class group may be half the size since the class group of integral binary quadratic forms is isomorphic to the narrow class group of $\mathbf{Q}(\sqrt{d})$.

Example of a Non-Trivial Class Group

The quadratic integer ring $R = \mathbf{Z}[\sqrt{-5}]$ is the ring of integers of $\mathbf{Q}(\sqrt{-5})$. It does *not* possess unique factorization; in fact the class group of *R* is cyclic of order 2. Indeed, the ideal

$J = (2, 1 + \sqrt{-5})$

is not principal, which can be proved by contradiction as follows. R has a norm function $N(a+b\sqrt{-5}) = a^2 + 5b^2$, which satisfies $N(uv) = N(u)N(v)$, and $N(u) = 1$ if and only if u is a unit in R. First of all, $J \neq R$, because the quotient ring of R modulo the ideal $(1+\sqrt{-5})$ is isomorphic to $\mathbf{Z}/6\mathbf{Z}$., so that the quotient ring of R modulo J is isomorphic to $\mathbf{Z}/2\mathbf{Z}$. If *J* were generated by an element *x* of *R*, then *x* would divide both 2 and $1 + \sqrt{-5}$. Then the norm $N(2) = 4$ and $N(1+\sqrt{5}) = 6$, so N(x) would divide 2. If $N(x) = 1$, then x is a unit, and J=R, a contradiction. But N(x) cannot be 2 either, because *R* has no elements of norm 2, because the Diophantine equation $b^2 + 5c^2 = 2$ has no solutions in integers, as it has no solutions modulo 5.

One also computes that $J^2 = (2)$, which is principal, so the class of *J* in the ideal class group has order two. Showing that there aren't any *other* ideal classes requires more effort.

The fact that this J is not principal is also related to the fact that the element 6 has two distinct factorisations into irreducibles:

$$6 = 2 \times 3 = (1 + \sqrt{-5}) \times (1 - \sqrt{-5}).$$

Connections to Class Field Theory

Class field theory is a branch of algebraic number theory which seeks to classify all the abelian extensions of a given algebraic number field, meaning Galois extensions with abelian Galois group. A particularly beautiful example is found in the Hilbert class field of a number field, which can be defined as the maximal unramified abelian extension of such a field. The Hilbert class field L of a number field K is unique and has the following properties:

- Every ideal of the ring of integers of K becomes principal in L, i.e., if I is an integral ideal of K then the image of I is a principal ideal in L.
- L is a Galois extension of K with Galois group isomorphic to the ideal class group of K.

Neither property is particularly easy to prove.

CLASS NUMBER PROBLEM

In mathematics, the Gauss class number problem (for imaginary quadratic fields), as usually understood, is to provide for each $n \geq 1$ a complete list of imaginary quadratic fields with class number n. It is named after Carl Friedrich Gauss. It can also be stated in terms of discriminants. There are related questions for real quadratic fields and the behavior as $d \to -\infty$.

The difficulty is in effective computation of bounds: for a given discriminant, it is easy to compute the class number, and there are several ineffective lower bounds on class number (meaning that they involve a constant that is not computed), but effective bounds (and explicit proofs of completeness of lists) are harder.

Gauss's Original Conjectures

The problems are posed in Gauss's Disquisitiones Arithmeticae of 1801 (Section V, Articles 303 and 304).

Gauss discusses imaginary quadratic fields in Article 303, stating the first two conjectures, and discusses real quadratic fields in Article 304, stating the third conjecture.

Gauss Conjecture (Class number tends to infinity)

$$h(d) \to \infty \text{ as } d \to -\infty.$$

Gauss Class Number Problem (Low Class Number Lists)

For given low class number (such as 1, 2, and 3), Gauss gives lists of imaginary quadratic fields with the given class number and believes them to be complete.

Infinitely Many Real Quadratic Fields with Class Number One

Gauss conjectures that there are infinitely many real quadratic fields with class number one.

The original Gauss class number problem for imaginary quadratic fields is significantly different and easier than the modern statement: he restricted to even discriminants, and allowed non-fundamental discriminants.

Status

Gauss Conjecture

Solved, Heilbronn, 1934.

Low Class Number Lists

Class number 1: solved, Baker (1966), Stark (1967), Heegner (1952).

Class number 2: solved, Baker (1971), Stark (1971)

Class number 3: solved, Oesterlé (1985)

Class numbers h up to 100: solved, Watkins 2004

Infinitely Many Real Quadratic Fields with Class Number One

For imaginary quadratic number fields, the (fundamental) discriminants of class number 1 are:

$$d = -3, -4, -7, -8, -11, -19, -43, -67, -163.$$

The non-fundamental discriminants of class number 1 are:

$$d = -12, -16, -27, -28.$$

Thus, the even discriminants of class number 1, fundamental and non-fundamental (Gauss's original question) are:

$$d = -4, -8, -12, -16, -28.$$

Modern Developments

In 1934, Hans Heilbronn proved the Gauss Conjecture. Equivalently, for any given class number, there are only finitely many imaginary quadratic number fields with that class number.

Also in 1934, Heilbronn and Edward Linfoot showed that there were at most 10 imaginary quadratic number fields with class number 1 (the 9 known ones, and at most one further). The result was ineffective: it did not give bounds on the size of the remaining field. In later developments, the case $n = 1$ was first discussed by Kurt Heegner, using modular forms and modular equations to show that no further such field could exist. This work was not initially accepted; only with later work of Harold Stark and Bryan Birch was the position clarified, and Heegner's work understood. Practically simultaneously, Alan Baker proved what we now know as Baker's theorem on linear forms in logarithms of algebraic numbers, which resolved the problem by a completely different method. The case $n = 2$ was tackled shortly afterwards, at least in principle, as an application of Baker's work.)

The complete list of imaginary quadratic fields with class number one is $\mathbf{Q}(\sqrt{k})$ with k one of

$$-1, -2, -3, -7, -11, -19, -43, -67, -163.$$

The general case awaited the discovery of Dorian Goldfeld that the class number problem could be connected to the L-functions of elliptic curves. This reduced the question, in principle, of effective determination, to one about establishing the existence of a multiple zero of such an L-function. This could be done on the basis of the later Gross-Zagier theorem. So at that point one could specify a finite calculation, the result of which would be a complete list for a given class number. In fact in practice such lists that are *probably* complete can be made by relatively simple methods; what is at issue is certainty. The cases up to $n = 100$ have now (2004) been done.

Real Quadratic Fields

The contrasting case of *real* quadratic fields is very different, and much less is known. That is because what enters the analytic formula for the class number is not h, the class number, on its own — but h log , where is a fundamental unit. This extra factor is hard to control. It may well be the case that class number 1 for real quadratic fields occurs infinitely often.

The Cohen-Lenstra heuristics are a set of more precise conjectures about the structure of class groups of quadratic fields. For real fields they predict that about 75.446% of the fields obtained by adjoining the square root of a prime will have class number 1, a result that agrees with computations.

PRINCIPAL IDEAL DOMAIN

In abstract algebra, a principal ideal domain, or PID, is an integral domain in which every ideal is principal, i.e., can be generated by a single element. More generally, a principal ideal ring is a nonzero commutative ring whose ideals are principal, although some authors (e.g., Bourbaki) refer to PIDs as principal rings. The distinction is that a principal ideal ring may have zero divisors whereas a principal ideal domain cannot.

Principal ideal domains are thus mathematical objects that behave somewhat like the integers, with respect to divisibility: any element of a PID has a unique decomposition into prime elements (so an analogue of the fundamental theorem of arithmetic holds); any two elements of a PID have a greatest common divisor (although it may not be possible to find it using the Euclidean algorithm). If x and y are elements of a PID without common divisors, then every element of the PID can be written in the form $ax + by$.

Principal ideal domains are noetherian, they are integrally closed, they are unique factorization domains and Dedekind domains. All Euclidean domains and all fields are principal ideal domains.

Principal ideal domains appear in the following chain of class inclusions:

commutative rings ⊃ integral domains ⊃ integrally closed domains ⊃ GCD domains ⊃ unique factorization domains ⊃ principal ideal domains ⊃ Euclidean domains ⊃ fields ⊃ finite fields

Examples

Examples include:

- K: any field,
- **Z**: the ring of integers,
- $K[x]$: rings of polynomials in one variable with coefficients in a field. (The converse is also true; that is, if $A[x]$ is a PID, then A is a field.) Furthermore, a ring of formal power series in one variable over a field is a PID since every ideal is of the form (x^k).
- **Z**[i]: the ring of Gaussian integers
- **Z**[ω] (where ω is a primitive cube root of 1): the Eisenstein integers

Examples of integral domains that are not PIDs:

- **Z**[x]: the ring of all polynomials with integer coefficients --- it is not principal because the ideal generated by 2 and X is an example of an ideal that cannot be generated by a single polynomial.
- $K[x,y]$: rings of polynomials in two variables. The ideal (x,y) is not principal.

Modules

Structure theorem for finitely generated modules over a principal ideal domain

The key result is the structure theorem: If R is a principal ideal domain, and M is a finitely generated R-module, then M is a direct sum of cyclic modules, i.e., modules with one generator. The cyclic modules are isomorphic to R/xR for some $x \in R$ (notice that x may be equal to 0, in which case R/xR is R)..

If M is a free module over a principal ideal domain R, then every submodule of M is again free. This does not hold for modules over arbitrary rings, as the example $(2, X) \subseteq \mathbb{Z}[X]$ of modules over $\mathbb{Z}[X]$ shows.

Properties

In a principal ideal domain, any two elements *a,b* have a greatest common divisor, which may be obtained as a generator of the ideal *(a,b)*.

All Euclidean domains are principal ideal domains, but the converse is not true. An example of a principal ideal domain that is not a Euclidean domain is the ring $\mathbb{Z}\left[(1+\sqrt{-19})/2\right]$ In this domain no q and r exist, with 0≤|r|<4, so that $(1+\sqrt{-19})=(4)q+r$, despite $1+\sqrt{-19}$ and 4 having a greatest common divisor of 2.

Every principal ideal domain is a unique factorization domain (UFD). The converse does not hold since for any UFD *K*, *K*[*X,Y*] (the rings of polynomials in 2 variables) is a UFD but is not a PID (to prove this look at the ideal generated by $\langle X,Y\rangle$ It is not the whole ring since it contains no polynomials of degree 0, but it cannot be generated by any one single element).

- Every principal ideal domain is Noetherian.
- In all unital rings, maximal ideals are prime. In principal ideal domains a near converse holds: every nonzero prime ideal is maximal.
- All principal ideal domains are integrally closed.

The previous three statements give the definition of a Dedekind domain, and hence every principal ideal domain is a Dedekind domain.

Let *A* be an integral domain. Then the following are equivalent.

- *A* is a PID.
- Every prime ideal of *A* is principal.
- *A* is a Dedekind domain that is a UFD.
- Every finitely generated ideal of *A* is principal (i.e., *A* is a Bézout domain) and *A* satisfies the ascending chain condition on principal ideals.
- *A* admits a Dedekind–Hasse norm.

A field norm is a Dedekind-Hasse norm; thus, (5) shows that a Euclidean domain is a PID. (4) compares to:

- An integral domain is a UFD if and only if it is a GCD domain (i.e., a domain where every two elements has a greatest common divisor) satisfying the ascending chain condition on principal ideals.

An integral domain is a Bézout domain if and only if any two elements in it have a gcd *that is a linear combination of the two.* A Bézout domain is thus a GCD domain, and (4) gives yet another proof that a PID is a UFD.

ALGEBRAIC K-THEORY

Algebraic *K*-theory is a subject area in mathematics with connections to geometry, topology, ring theory, and number theory. Geometric, algebraic, and arithmetic objects are assigned objects called *K*-groups. These are groups in the sense of abstract algebra. They contain detailed information about the original object but are notoriously difficult to compute; for example, an important outstanding problem is to compute the *K*-groups of the integers.

K-theory was invented in the late 1950s by Alexander Grothendieck in his study of intersection theory on algebraic varieties. Intersection theory is still a motivating force in the development of algebraic K-theory through its links with motivic cohomology and specifically Chow groups. The subject also includes classical number-theoretic topics like quadratic reciprocity and embeddings of number fields into the real numbers and complex numbers, as well as more modern concerns like the construction of higher regulators and special values of L-functions.

The lower K-groups were discovered first, in the sense that adequate descriptions of these groups in terms of other algebraic structures were found. For example, if F is a field, then $K_0(F)$ is isomorphic to the integers $\mathbf{Z}$ and is closely related to the notion of vector space dimension. For a commutative ring R, $K_0(R)$ is the Picard group of R, and when R is the ring of integers in a number field, this generalizes the classical construction of the class group. The group $K_1(R)$ is closely related to the group of units $R^\times$, and if R is a field, it is exactly the group of units. For a number field F, $K_2(F)$ is related to class field theory, the Hilbert symbol, and the solvability of quadratic equations over completions. In contrast, finding the correct definition of the higher K-groups of rings was a difficult achievement of Daniel Quillen, and many of the basic facts about the higher K-groups of algebraic varieties were not known until the work of Robert Thomason.

History

The history of K-theory was detailed by Weibel.

K_0, K_1, and K_2

In the 19th century, Bernhard Riemann and his student Gustav Roch proved what is now known as the Riemann–Roch theorem. If X is a Riemann surface, then the sets of meromorphic functions and meromorphic differential forms on X form vector spaces. A line bundle on X determines subspaces of these vector spaces, and if X is projective, then these subspaces are finite dimensional. The Riemann–Roch theorem states that the difference in dimensions between these subspaces is equal to the degree of the line bundle (a measure of twistedness) plus one minus the genus of X. In the mid-20th century, the Riemann–Roch theorem was generalized by Friedrich Hirzebruch to all algebraic varieties. In Hirzebruch's formulation, the Hirzebruch–Riemann–Roch theorem, the theorem became a statement about Euler characteristics: The Euler characteristic of a vector bundle on an algebraic variety (which is the alternating sum of the dimensions of its cohomology groups) equals the Euler characteristic of the trivial bundle plus a correction factor coming from characteristic classes of the vector bundle. This is a generalization because on a projective Riemann surface, the Euler characteristic of a line bundle equals the difference in dimensions mentioned previously, the Euler characteristic of the trivial bundle is one minus the genus, and the only non-trivial characteristic class is the degree.

The subject of K-theory takes its name from a 1957 construction of Alexander Grothendieck which appeared in the Grothendieck–Riemann–Roch theorem, his generalization of Hirzebruch's theorem. Let X be a smooth algebraic variety. To each vector bundle on X, Grothendieck associates an invariant, its *class*. The set of all classes on X was called $K(X)$ from the German *Klasse*. By definition, $K(X)$ is a quotient of the free abelian group on isomorphism classes of vector bundles on X, and so it is an abelian group. If the basis element corresponding to a vector bundle V is denoted $[V]$, then for each short exact sequence of vector bundles:

$$0 \to V' \to V \to V'' \to 0,$$

Grothendieck imposed the relation $[V] = [V'] + [V'']$. These generators and relations define $K(X)$, and they imply that it is the universal way to assign invariants to vector bundles in a way compatible with exact sequences.

Grothendieck took the perspective that the Riemann–Roch theorem is a statement about morphisms of varieties, not the varieties themselves. He proved that there is a homomorphism from $K(X)$ to the Chow groups of X coming from the Chern character and Todd class of X. Additionally, he proved that a proper morphism $f : X \to Y$ to a smooth variety Y determines a homomorphism $f_* : K(X) \to K(Y)$ called the *pushforward*. This gives two ways of determining an element in the Chow group of Y from a vector bundle on X: Starting from X, one can first compute the pushforward in K-theory and then apply the Chern character and Todd class of Y, or one can first apply the Chern character and Todd class of X and then compute the pushforward for Chow groups. The Grothendieck–Riemann–Roch theorem says that these are equal. When Y is a point, a vector bundle is a vector space, the class of a vector space is its dimension, and the Grothendieck–Riemann–Roch theorem specializes to Hirzebruch's theorem.

The group $K(X)$ is now known as $K_0(X)$. Upon replacing vector bundles by projective modules, K_0 also became defined for non-commutative rings, where it had applications to group representations. Atiyah and Hirzebruch quickly transported Grothendieck's construction to topology and used it to define topological K-theory. Topological K-theory was one of the first examples of an extraordinary cohomology theory: It associates to each topological space X (satisfying some mild technical constraints) a sequence of groups $K_n(X)$ which satisfy all the Eilenberg–Steenrod axioms except the normalization axiom. The setting of algebraic varieties, however, is much more rigid, and the flexible constructions used in topology were not available. While the group K_0 seemed to satisfy the necessary properties to be the beginning of a cohomology theory of algebraic varieties and of non-commutative rings, there was no clear definition of the higher $K_n(X)$. Even as such definitions were developed, technical issues surrounding restriction and gluing usually forced K_n to be defined only for rings, not for varieties.

While it was not initially known, a group related to K_1 had already been introduced in another context. Henri Poincaré had attempted to define the Betti numbers of a manifold in terms of a triangulation. His methods, however, had a serious gap: Poincaré could not prove that two triangulations of a manifold always yielded the same Betti numbers. It was clearly true that Betti numbers were unchanged by subdividing the triangulation, and therefore it was clear that any two triangulations that shared a common subdivision had the same Betti numbers. What was not known was that any two triangulations admitted a common subdivision. This hypothesis became a conjecture known as the *Hauptvermutung* (roughly "main conjecture"). The fact that triangulations were stable under subdivision led J.H.C. Whitehead to introduce the notion of simple homotopy type. A simple homotopy equivalence is defined in terms of adding simplices or cells to a simplicial complex or cell complex in such a way that each additional simplex or cell deformation retracts into a subdivision of the old space. Part of the motivation for this definition is that a subdivision of a triangulation is simple homotopy equivalent to the original triangulation, and therefore two triangulations that share a common subdivision must be simple homotopy equivalent. Whitehead proved that simple homotopy equivalence is a finer invariant than homotopy equivalence by introducing an invariant called the *torsion*. The torsion of a homotopy equivalence takes values in a group now called the *Whitehead group* and denoted $Wh(\pi)$, where π is the fundamental group of the two complexes. Whitehead found examples of non-trivial torsion and thereby proved that some homotopy equivalences were not simple. The Whitehead group was later discovered to be a quotient of $K_1(\mathbf{Z}\pi)$, where $\mathbf{Z}\pi$ is the integral group ring of π. Later John Milnor used Reidemeister torsion, an invariant related to Whitehead torsion, to disprove the Hauptvermutung.

The first adequate definition of K_1 of a ring was made by Hyman Bass and Stephen Schanuel. In topological *K*-theory, K_1 is defined using vector bundles on a suspension of the space. All such vector bundles come from the clutching construction, where two trivial vector bundles on two halves of a space are glued along a common strip of the space. This gluing data is expressed using the general linear group, but elements of that group coming from elementary matrices (matrices corresponding to elementary row or column operations) define equivalent gluings. Motivated by this, the Bass–Schanuel definition of K_1 of a ring R is $GL(R) / E(R)$, where $GL(R)$ is the infinite general linear group (the union of all $GL_n(R)$) and $E(R)$ is the subgroup of elementary matrices. They also provided a definition of K_0 of a homomorphism of rings and proved that K_0 and K_1 could be fit together into an exact sequence similar to the relative homology exact sequence.

Work in *K*-theory from this period culminated in Bass' book *Algebraic* K-*theory*. In addition to providing a coherent exposition of the results then known, Bass improved many of the statements of the theorems. Of particular note is that Bass, building on his earlier work with Murthy, provided the first proof of what is now known as the fundamental theorem of algebraic *K*-theory. This is a four-term exact sequence relating K_0 of a ring R to K_1 of R, the polynomial ring $R[t]$, and the localization $R[t, t^{-1}]$. Bass recognized that this theorem provided a description of K_0 entirely in terms of K_1. By applying this description recursively, he produced negative *K*-groups $K_{-n}(R)$. In independent work, Max Karoubi gave another definition of negative *K*-groups for certain categories and proved that his definitions yielded that same groups as those of Bass.

The next major development in the subject came with the definition of K_2. Steinberg studied the universal central extensions of a Chevalley group over a field and gave an explicit presentation of this group in terms of generators and relations. In the case of the group $E_n(k)$ of elementary matrices, the universal central extension is now written $St_n(k)$ and called the *Steinberg group*. In the spring of 1967, John Milnor defined $K_2(R)$ to be the kernel of the homomorphism $St(R) \to E(R)$. The group K_2 further extended some of the exact sequences known for K_1 and K_0, and it had striking applications to number theory. Matsumoto's 1968 thesis showed that for a field F, $K_2(F)$ was isomorphic to:

$$F^\times \otimes_{\mathbf{Z}} F^\times / \langle x \otimes (1-x) : x \in F \setminus \{0,1\} \rangle.$$

This relation is also satisfied by the Hilbert symbol, which expresses the solvability of quadratic equations over local fields. In particular, John Tate was able to prove that $K_2(\mathbf{Q})$ is essentially structured around the law of quadratic reciprocity.

Higher K-groups

In the late 1960s and early 1970s, several definitions of higher *K*-theory were proposed. Swan and Gersten both produced definitions of K_n for all *n*, and Gersten proved that his and Swan's theories were equivalent, but the two theories were not known to satisfy all the expected properties. Nobile and Villamayor also proposed a definition of higher *K*-groups. Karoubi and Villamayor defined well-behaved *K*-groups for all *n*, but their equivalent of K_1 was sometimes a proper quotient of the Bass–Schanuel K_1. Their *K*-groups are now called KV_n and are related to homotopy-invariant modifications of *K*-theory.

Inspired in part by Matsumoto's theorem, Milnor made a definition of the higher *K*-groups of a field. He referred to his definition as "purely *ad hoc*", and it neither appeared to generalize to all rings nor did it appear to be the correct definition of the higher *K*-theory of fields. Much later, it was discovered by Nesterenko and Suslin and by Totaro that Milnor *K*-theory is actually a direct summand of the true *K*-theory of the field. Specifically, *K*-groups have a filtration called the *weight filtration*, and the Milnor *K*-theory of a field is the highest weight-graded piece of the *K*-theory. Additionally, Thomason discovered that there is no analog of Milnor *K*-theory for a general variety.

The first definition of higher *K*-theory to be widely accepted was Daniel Quillen's. As part of Quillen's work on the Adams conjecture in topology, he had constructed maps from the classifying spaces $BGL(\mathbf{F}_q)$ to the homotopy fiber of $\psi^q - 1$, where ψ^q is the *q*th Adams operation acting on the classifying space *BU*. This map is acyclic, and after modifying $BGL(\mathbf{F}_q)$ slightly to produce a new space $BGL(\mathbf{F}_q)^+$, the map became a homotopy equivalence. This modification was called the plus construction. The Adams operations had been known to be related to Chern classes and to *K*-theory since the work of Grothendieck, and so Quillen was led to define the *K*-theory of *R* as the homotopy groups of $BGL(R)^+$. Not only did this recover K_1 and K_2, the relation of *K*-theory to the Adams operations allowed Quillen to compute the *K*-groups of finite fields.

The classifying space *BGL* is connected, so Quillen's definition failed to give the correct value for K_0. Additionally, it did not give any negative *K*-groups. Since K_0 had a known and accepted definition it was possible to sidestep this difficulty, but it remained technically awkward. Conceptually, the problem was that the definition sprung from *GL*, which was classically the source of K_1. Because *GL* knows only about gluing vector bundles, not about the vector bundles themselves, it was impossible for it to describe K_0.

Inspired by conversations with Quillen, Segal soon introduced another approach to constructing algebraic *K*-theory under the name of Γ-objects. Segal's approach is a homotopy analog of Grothendieck's construction of K_0. Where Grothendieck worked with isomorphism classes of bundles, Segal worked with the bundles themselves and used isomorphisms of the bundles as part of his data. This results in a spectrum whose homotopy groups are the higher *K*-groups (including K_0). However, Segal's approach was only able to impose relations for split exact sequences, not general exact sequences. In the category of projective modules over a ring, every short exact sequence splits, and so

Γ-objects could be used to define the *K*-theory of a ring. However, there are non-split short exact sequences in the category of vector bundles on a variety and in the category of all modules over a ring, so Segal's approach did not apply to all cases of interest.

In the spring of 1972, Quillen found another approach to the construction of higher *K*-theory which was to prove enormously successful. This new definition began with an exact category, a category satisfying certain formal properties similar to, but slightly weaker than, the properties satisfied by a category of modules or vector bundles. From this he constructed an auxiliary category using a new device called his "*Q*-construction." Like Segal's Γ-objects, the *Q*-construction has its roots in Grothendieck's definition of K_0. Unlike Grothendieck's definition, however, the *Q*-construction builds a category, not an abelian group, and unlike Segal's Γ-objects, the *Q*-construction works directly with short exact sequences. If *C* is an abelian category, then *QC* is a category with the same objects as *C* but whose morphisms are defined in terms of short exact sequences in *C*. The *K*-groups of the exact category are the homotopy groups of ΩBQC, the loop space of the geometric realization (taking the loop space corrects the indexing). Quillen additionally proved his "+ = *Q* theorem" that his two definitions of *K*-theory agreed with each other. This yielded the correct K_0 and led to simpler proofs, but still did not yield any negative *K*-groups.

All abelian categories are exact categories, but not all exact categories are abelian. Because Quillen was able to work in this more general situation, he was able to use exact categories as tools in his proofs. This technique allowed him to prove many of the basic theorems of algebraic *K*-theory. Additionally, it was possible to prove that the earlier definitions of Swan and Gersten were equivalent to Quillen's under certain conditions.

K-theory now appeared to be a homology theory for rings and a cohomology theory for varieties. However, many of its basic theorems carried the hypothesis that the ring or variety in question was regular. One of the basic expected relations was a long exact sequence (called the "localization sequence") relating the *K*-theory of a variety *X* and an open subset *U*. Quillen was unable to prove the existence of the localization sequence in full generality. He was, however, able to prove its existence for a related theory called *G*-theory (or sometimes *K*′-theory). *G*-theory had been defined early in the development of the subject by Grothendieck. Grothendieck defined $G_0(X)$ for a variety *X* to be the free abelian group on isomorphism classes of coherent sheaves on *X*, modulo relations coming from exact sequences of coherent sheaves. In the categorical framework adopted by later authors, the *K*-theory of a variety is the *K*-theory of its category of vector bundles, while its *G*-theory is the *K*-theory of its category of coherent sheaves. Not only could Quillen prove the existence of a localization exact sequence for *G*-theory, he could prove that for a regular ring or variety, *K*-theory equaled *G*-theory, and therefore *K*-theory of regular varieties had a localization exact sequence. Since this sequence was fundamental to many of the facts in the subject, regularity hypotheses pervaded early work on higher *K*-theory.

Applications of Algebraic K-theory in Topology

The earliest application of algebraic *K*-theory to topology was Whitehead's construction of Whitehead torsion. A closely related construction was found by C. T. C. Wall in 1963. Wall found that a space π dominated by a finite complex has a generalized Euler characteristic taking values in a quotient of $K_0(\mathbf{Z}\pi)$, where π is the fundamental group of the space. This invariant is called *Wall's finiteness obstruction* because X is homotopy equivalent to a finite complex if and only if the invariant vanishes. Laurent Siebenmann in his thesis found an invariant similar to Wall's that gives an obstruction to an open manifold being the interior of a compact manifold with boundary. If two manifolds with boundary M and N have isomorphic interiors (in TOP, PL, or DIFF as appropriate), then the isomorphism between them defines an h-cobordism between M and N.

Whitehead torsion was eventually reinterpreted in a more directly K-theoretic way. This reinterpretation happened through the study of h-cobordisms. Two n-dimensional manifolds M and N are h-cobordant if there exists an $(n + 1)$-dimensional manifold with boundary W whose boundary is the disjoint union of M and N and for which the inclusions of M and N into W are homotopy equivalences (in the categories TOP, PL, or DIFF). Stephen Smale's h-cobordism theorem asserted that if $n \geq 5$, W is compact, and M, N, and W are simply connected, then W is isomorphic to the cylinder $M \times [0, 1]$ (in TOP, PL, or DIFF as appropriate). This theorem proved the Poincaré conjecture for $n \geq 5$.

If M and N are not assumed to be simply connected, then an h-cobordism need not be a cylinder. The s-cobordism theorem, due independently to Mazur, Stallings, and Barden,explains the general situation: An h-cobordism is a cylinder if and only if the Whitehead torsion of the inclusion $M \subset W$ vanishes. This generalizes the h-cobordism theorem because the simple connectedness hypotheses imply that the relevant Whitehead group is trivial. In fact the s-cobordism theorem implies that there is a bijective correspondence between isomorphism classes of h-cobordisms and elements of the Whitehead group.

An obvious question associated with the existence of h-cobordisms is their uniqueness. The natural notion of equivalence is isotopy. Jean Cerf proved that for simply connected smooth manifolds M of dimension at least 5, isotopy of h-cobordisms is the same as a weaker notion called pseudo-isotopy. Hatcher and Wagoner studied the components of the space of pseudo-isotopies and related it to a quotient of $K_2(\mathbf{Z}\pi)$.

The proper context for the s-cobordism theorem is the classifying space of h-cobordisms. If M is a CAT manifold, then $H^{\mathrm{CAT}}(M)$ is a space that classifies bundles of h-cobordisms on M. The s-cobordism theorem can be reinterpreted as the statement that the set of connected components of this space is the Whitehead group of $\pi_1(M)$. This space contains strictly more information than the Whitehead group; for example, the connected component of the trivial cobordism describes the possible cylinders on M and in particular is the obstruction to the uniqueness of a homotopy between a manifold and $M \times [0, 1]$. Consideration of these questions led Waldhausen to introduced his *algebraic* K-*theory of spaces*. The algebraic K-theory of M is a space $A(M)$ which is defined so that it plays essentially the same role for higher K-groups as $K_1(\mathbf{Z}\pi_1(M))$ does for M. In particular, Waldhausen showed that there is a map from $A(M)$ to a space Wh(M) which generalizes the map $K_1(\mathbf{Z}\pi_1(M)) \rightarrow \mathrm{Wh}(\pi_1(M))$ and whose homotopy fiber is a homology theory.

In order to fully develop A-theory, Waldhausen made significant technical advances in the foundations of K-theory. Waldhausen introduced Waldhausen categories, and for a Waldhausen category C he introduced a simplicial category $S.C$ (the S is for Segal) defined in terms of chains of cofibrations in C. This freed the foundations of K-theory from the need to invoke analogs of exact sequences.

Algebraic Topology and Algebraic Geometry in Algebraic K-theory

Quillen suggested to his student Kenneth Brown that it might be possible to create a theory of sheaves of spectra of which *K*-theory would provide an example. The sheaf of *K*-theory spectra would, to each open subset of a variety, associate the *K*-theory of that open subset. Brown developed such a theory for his thesis. Simultaneously, Gersten had the same idea. At a Seattle conference in autumn of 1972, they together discovered a spectral sequence converging from the sheaf cohomology of $\mathcal{K}_n$, the sheaf of K_n-groups on *X*, to the *K*-group of the total space. This is now called the Brown–Gersten spectral sequence.

Spencer Bloch, influenced by Gersten's work on sheaves of *K*-groups, proved that on a regular surface, the cohomology group $H^2(X, \mathcal{K}_2)$ is isomorphic to the Chow group $CH^2(X)$ of codimension 2 cycles on *X*. Inspired by this, Gersten conjectured that for a regular local ring *R* with fraction field *F*, $K_n(R)$ injects into $K_n(F)$ for all *n*. Soon Quillen proved that this is true when *R* contains a field, and using this he proved that $H^p(X, \mathcal{K}_p) \cong \mathrm{CH}^p(X)$ for all *p*. This is known as *Bloch's formula*. While progress has been made on Gersten's conjecture since then, the general case remains open.

Lichtenbaum conjectured that special values of the zeta function of a number field could be expressed in terms of the *K*-groups of the ring of integers of the field. These special values were known to be related to the etale cohomology of the ring of integers. Quillen therefore generalized Lichtenbaum's conjecture, predicting the existence of a spectral sequence like the Atiyah–Hirzebruch spectral sequence in topological *K*-theory. Quillen's proposed spectral sequence would start from the etale cohomology of a ring *R* and, in high enough degrees and after completing at a prime l invertible in *R*, abut to the l-adic completion of the *K*-theory of *R*. In the case studied by Lichtenbaum, the spectral sequence would degenerate, yielding Lichtenbaum's conjecture.

The necessity of localizing at a prime l suggested to Browder that there should be a variant of *K*-theory with finite coefficients. He introduced *K*-theory groups $K_n(R; \mathbf{Z}/l\mathbf{Z})$ which were $\mathbf{Z}/l\mathbf{Z}$-vector spaces, and he found an analog of the Bott element in topological *K*-theory. Soule used this theory to construct "etale Chern classes", an analog of topological Chern classes which took elements of algebraic *K*-theory to classes in etale cohomology. Unlike algebraic *K*-theory, etale cohomology is highly computable, so etale Chern classes provided an effective tool for detecting the existence of elements in *K*-theory. Dwyer and Friedlander then invented an analog of *K*-theory for the etale topology called etale *K*-theory. For varieties defined over the complex numbers, etale *K*-theory is isomorphic to topological *K*-theory. Moreover, etale *K*-theory admitted a spectral sequence similar to the one conjectured by Quillen. Thomason proved around 1980 that after inverting the Bott element, algebraic *K*-theory with finite coefficients became isomorphic to etale *K*-theory.

Throughout the 1970s and early 1980s, *K*-theory on singular varieties still lacked adequate foundations. While it was believed that Quillen's *K*-theory gave the correct groups, it was not known that these groups had all of the envisaged properties. For this, algebraic *K*-theory had to be reformulated. This was done by Thomason in a lengthy monograph which he co-credited to his dead friend Thomas Trobaugh, who he said gave him a key idea in a dream. Thomason combined Waldhausen's construction of *K*-theory with the foundations of intersection theory described in volume six of Grothendieck's Séminaire de Géométrie Algébrique du Bois Marie. There, K_0 was described in terms of complexes of sheaves on algebraic varieties. Thomason discovered that if one worked with in derived category of sheaves, there was a simple description of when a complex of sheaves could be extended from an open subset of a variety to the whole variety. By applying Waldhausen's construction of K-theory to derived categories, Thomason was able to prove that algebraic K-theory had all the expected properties of a cohomology theory.

In 1976, Keith Dennis discovered an entirely novel technique for computing K-theory based on Hochschild homology. This was based around the existence of the Dennis trace map, a homomorphism from K-theory to Hochschild homology. While the Dennis trace map seemed to be successful for calculations of K-theory with finite coefficients, it was less successful for rational calculations. Goodwillie, motivated by his "calculus of functors", conjectured the existence of a theory intermediate to K-theory and Hochschild homology. He called this theory topological Hochschild homology because its ground ring should be the sphere spectrum (considered as a ring whose operations are defined only up to homotopy). In the mid-80s, Bokstedt gave a definition of topological Hochschild homology that satisfied nearly all of Goodwillie's conjectural properties, and this made possible further computations of K-groups. Bokstedt's version of the Dennis trace map was a transformation of spectra $K \to THH$. This transformation factored through the fixed points of a circle action on THH, which suggested a relationship with cyclic homology. In the course of proving an algebraic K-theory analog of the Novikov conjecture, Bokstedt, Hsiang, and Madsen introduced topological cyclic homology, which bore the same relationship to topological Hochschild homology as cyclic homology did to Hochschild homology. The Dennis trace map to topological Hochschild homology facts through topological cyclic homology, providing an even more detailed tool for calculations. In 1996, Dundas, Goodwillie, and McCarthy proved that topological cyclic homology has in a precise sense the same local structure as algebraic K-theory, so that if a calculation in K-theory or topological cyclic homology is possible, then many other "nearby" calculations follow.

Lower K-Groups

The lower K-groups were discovered first, and given various ad hoc descriptions, which remain useful. Throughout, let A be a ring.

K_0

The functor K_0 takes a ring A to the Grothendieck group of the set of isomorphism classes of its finitely generated projective modules, regarded as a monoid under direct sum. Any ring homomorphism $A \to B$ gives a map $K_0(A) \to K_0(B)$ by mapping (the class of) a projective A-module M to $M \otimes_A B$, making K_0 a covariant functor.

If the ring A is commutative, we can define a subgroup of $K_0(A)$ as the set

$$\tilde{K}_0(A) = \bigcap_{\mathfrak{p} \text{ prime ideal of } A} \operatorname{Ker} \dim_{\mathfrak{p}},$$

where :

$$\dim_{\mathfrak{p}} : K_0(A) \to \mathbf{Z}$$

is the map sending every (class of a) finitely generated projective A-module M to the rank of the free $A_{\mathfrak{p}}$-module $M_{\mathfrak{p}}$ (this module is indeed free, as any finitely generated projective module over a local ring is free). This subgroup $\tilde{K}_0(A)$ is known as the *reduced zeroth K-theory* of A.

If B is a ring without an identity element, we can extend the definition of K_0 as follows. Let $A = B \oplus \mathbf{Z}$ be the extension of B to a ring with unity obtaining by adjoining an identity element (0,1). There is a short exact sequence $B \to A \to \mathbf{Z}$ and we define $K_0(B)$ to be the kernel of the corresponding map $K_0(A) \to K_0(\mathbf{Z}) = \mathbf{Z}$.

Examples

- (Projective) modules over a field k are vector spaces and $K_0(k)$ is isomorphic to $\mathbf{Z}$, by dimension.
- Finitely generated projective modules over a local ring A are free and so in this case again $K_0(A)$ is isomorphic to $\mathbf{Z}$, by rank.
- For A a Dedekind domain,

$K_0(A) = \mathrm{Pic}(A) \oplus \mathbf{Z}$,

where Pic(A) is the Picard group of A, and similarly the reduced K-theory is given by

$$\tilde{K}_0(A) = \operatorname{Pic} A.$$

An algebro-geometric variant of this construction is applied to the category of algebraic varieties; it associates with a given algebraic variety X the Grothendieck's K-group of the category of locally free sheaves (or coherent sheaves) on X. Given a compact topological space X, the topological K-theory $K^{top}(X)$ of (real) vector bundles over X coincides with K_0 of the ring of continuous real-valued functions on X.

Relative K_0

Let I be an ideal of A and define the "double" to be a subring of the Cartesian product $A \times A$:

$$D(A, I) = \{(x, y) \in A \times A : x - y \in I\}.$$

The *relative K-group* is defined in terms of the "double"

$$K_0(A, I) = \ker(K_0(D(A, I)) \to K_0(A)).$$

where the map is induced by projection along the first factor.

The relative $K_0(A,I)$ is isomorphic to $K_0(I)$, regarding I as a ring without identity. The independence from A is an analogue of the Excision theorem in homology.

K_0 as a Ring

If A is a commutative ring, then the tensor product of projective modules is again projective, and so tensor product induces a multiplication turning K_0 into a commutative ring with the class $[A]$ as identity. The exterior product similarly induces a λ-ring structure. The Picard group embeds as a subgroup of the group of units $K_0(A)^*$.

K_1

Hyman Bass provided this definition, which generalizes the group of units of a ring: $K_1(A)$ is the abelianization of the infinite general linear group:

$$K_1(A) = \mathrm{GL}(A)^{\mathrm{ab}} = \mathrm{GL}(A)/[\mathrm{GL}(A), \mathrm{GL}(A)]$$

Here

$$\mathrm{GL}(A) = \operatorname{colim} \mathrm{GL}(n, A)$$

is the direct limit of the GL(n), which embeds in GL(n + 1) as the upper left block matrix, and the commutator subgroup agrees with the group generated by elementary matrices $E(A)$ = [GL(A), GL(A)], by Whitehead's lemma. Indeed, the group GL(A)/E(A) was first defined and studied by Whitehead, and is called the Whitehead group of the ring A.

Relative K_1

The *relative K-group* is defined in terms of the "double"

$$K_1(A, I) = \ker(K_1(D(A, I)) \to K_1(A)) .$$

There is a natural exact sequence

$$K_1(A, I) \to K_1(A) \to K_1(A/I) \to K_0(A, I) \to K_0(A) \to K_0(A/I) .$$

Commutative Rings and Fields

For A a commutative ring, one can define a determinant det: GL(A) → A^* to the group of units of A, which vanishes on E(A) and thus descends to a map det: $K_1(A) \to A^*$. As E(A) ◁ SL(A), one can also define the special Whitehead group $SK_1(A)$:= SL(A)/E(A). This map splits via the map $A^* \to$ GL(1, A) → $K_1(A)$ (unit in the upper left corner), and hence is onto, and has the special Whitehead group as kernel, yielding the split short exact sequence:

$$1 \to SK_1(A) \to K_1(A) \to A^* \to 1,$$

which is a quotient of the usual split short exact sequence defining the special linear group, namely

$$1 \to \mathbf{SL}(A) \to \mathbf{GL}(A) \to A^* \to 1.$$

The determinant is split by including the group of units $A^* = \mathrm{GL}_1(A)$ into the general linear group $\mathrm{GL}(A)$, so $K_1(A)$ splits as the direct sum of the group of units and the special Whitehead group: $K_1(A) \cong A^* \oplus \mathrm{SK}_1(A)$.

When A is a Euclidean domain (e.g. a field, or the integers) $\mathrm{SK}_1(A)$ vanishes, and the determinant map is an isomorphism from $\mathrm{K}_1(A)$ to A^*. This is *false* in general for PIDs, thus providing one of the rare mathematical features of Euclidean domains that do not generalize to all PIDs. An explicit PID such that SK_1 is nonzero was given by Ischebeck in 1980 and by Grayson in 1981. If A is a Dedekind domain whose quotient field is an algebraic number field (a finite extension of the rationals) then Milnor (1971, corollary 16.3) shows that $\mathrm{SK}_1(A)$ vanishes.

The vanishing of SK_1 can be interpreted as saying that K_1 is generated by the image of GL_1 in GL. When this fails, one can ask whether K_1 is generated by the image of GL_2. For a Dedekind domain, this is the case: indeed, K_1 is generated by the images of GL_1 and SL_2 in GL. The subgroup of SK_1 generated by SL_2 may be studied by Mennicke symbols. For Dedekind domains with all quotients by maximal ideals finite, SK_1 is a torsion group.

For a non-commutative ring, the determinant cannot in general be defined, but the map $\mathrm{GL}(A) \to K_1(A)$ is a generalization of the determinant.

Central Simple Algebras

In the case of a central simple algebra A over a field F, the reduced norm provides a generalisation of the determinant giving a map $K_1(A) \to F^*$ and $\mathrm{SK}_1(A)$ may be defined as the kernel. Wang's theorem states that if A has prime degree then $\mathrm{SK}_1(A)$ is trivial, and this may be extended to square-free degree. Wang also showed that $\mathrm{SK}_1(A)$ is trivial for any central simple algebra over a number field, but Platonov has given examples of algebras of degree prime squared for which $\mathrm{SK}_1(A)$ is non-trivial.

K_2

John Milnor found the right definition of K_2: it is the center of the Steinberg group $\mathrm{St}(A)$ of A.

It can also be defined as the kernel of the map

$$\varphi\colon \mathrm{St}(A) \to \mathrm{GL}(A),$$

or as the Schur multiplier of the group of elementary matrices.

For a field, K_2 is determined by Steinberg symbols: this leads to Matsumoto's theorem.

One can compute that K_2 is zero for any finite field. The computation of $\mathrm{K}_2(\mathbf{Q})$ is complicated: Tate proved

$$K_2(\mathbf{Q}) = (\mathbf{Z}/4)^* \times \prod_{p \text{ odd prime}} (\mathbf{Z}/p)^*$$

and remarked that the proof followed Gauss's first proof of the Law of Quadratic Reciprocity.

For non-Archimedean local fields, the group $K_2(F)$ is the direct sum of a finite cyclic group of order m, say, and a divisible group $K_2(F)^m$.

We have $K_2(\mathbf{Z}) = \mathbf{Z}/2$, and in general K_2 is finite for the ring of integers of a number field.

We further have $K_2(\mathbf{Z}/n) = \mathbf{Z}/2$ if n is divisible by 4, and otherwise zero.

Matsumoto's Theorem

Matsumoto's theorem states that for a field k, the second K-group is given by

$$K_2(k) = k^\times \otimes_{\mathbf{Z}} k^\times / \langle a \otimes (1-a) \mid a \neq 0, 1 \rangle.$$

Matsumoto's original theorem is even more general: For any root system, it gives a presentation for the unstable K-theory. This presentation is different from the one given here only for symplectic root systems. For non-symplectic root systems, the unstable second K-group with respect to the root system is exactly the stable K-group for GL(A). Unstable second K-groups (in this context) are defined by taking the kernel of the universal central extension of the Chevalley group of universal type for a given root system. This construction yields the kernel of the Steinberg extension for the root systems A_n ($n > 1$) and, in the limit, stable second K-groups.

Long Exact Sequences

If A is a Dedekind domain with field of fractions F then there is a long exact sequence

$$K_2F \to \oplus_{\mathbf{p}} K_1 A/\mathbf{p} \to K_1 A \to K_1 F \to \oplus_{\mathbf{p}} K_0 A/\mathbf{p} \to K_0 A \to K_0 F \to 0$$

where p runs over all prime ideals of A.

There is also an extension of the exact sequence for relative K_1 and K_0:

$$K_2(A) \to K_2(A/I) \to K_1(A,I) \to K_1(A) \cdots.$$

Pairing

There is a pairing on K_1 with values in K_2. Given commuting matrices X and Y over A, take elements x and y in the Steinberg group with X,Y as images. The commutator $xyx^{-1}y^{-1}$ is an element of K_2. The map is not always surjective.

Milnor *K*-Theory

The above expression for K_2 of a field k led Milnor to the following definition of "higher" *K*-groups by

$$K_*^M(k) := T^*(k^\times)/(a \otimes (1-a)),$$

thus as graded parts of a quotient of the tensor algebra of the multiplicative group $k^\times$ by the two-sided ideal, generated by the

$$\{a \otimes (1-a) : a \neq 0, 1\}.$$

For $n = 0,1,2$ these coincide with those below, but for $n \geqq 3$ they differ in general. For example, we have $KMn(F_q) = 0$ for $n \geqq 2$ but K_nF_q is nonzero for odd n (see below).

The tensor product on the tensor algebra induces a product $K_m \times K_n \to K_{m+n}$ making $K_*^M(F)$ a graded ring which is graded-commutative.

The images of elements $a_1 \otimes \cdots \otimes a_n$ in $K_n^M(k)$ are termed *symbols*, denoted $\{a_1, \ldots, a_n\}$. For integer *m* invertible in *k* there is a map

$$\partial : k^* \to H^1(k, \mu_m)$$

where μ_m denotes the group of *m*-th roots of unity in some separable extension of *k*. This extends to

$$\partial^n : k^* \times \cdots \times k^* \to H^n\left(k, \mu_m^{\otimes n}\right)$$

satisfying the defining relations of the Milnor K-group. Hence ∂^n may be regarded as a map on $K_n^M(k)$, called the *Galois symbol* map.

The relation between étale (or Galois) cohomology of the field and Milnor K-theory modulo 2 is the Milnor conjecture, proven by Vladimir Voevodsky. The analogous statement for odd primes is the Bloch-Kato conjecture, proved by Voevodsky, Rost, and others.

Higher *K*-Theory

The accepted definitions of higher *K*-groups were given by Quillen (1973), after a few years during which several incompatible definitions were suggested. The object of the program was to find definitions of **K**(R) and **K**(R,I) in terms of classifying spaces so that $R \Rightarrow$ **K**(R) and (R,I) $\Rightarrow$ **K**(R,I) are functors into a homotopy category of spaces and the long exact sequence for relative K-groups arises as the long exact homotopy sequence of a fibration **K**(R,I) → **K**(R) → **K**(R/I).

Quillen gave two constructions, the "plus-construction" and the "*Q*-construction", the latter subsequently modified in different ways. The two constructions yield the same K-groups.

The +-Construction

One possible definition of higher algebraic *K*-theory of rings was given by Quillen

$$K_n(R) = \pi_n(BGL(R)^+),$$

Here π_n is a homotopy group, GL(R) is the direct limit of the general linear groups over R for the size of the matrix tending to infinity, B is the classifying space construction of homotopy theory, and the $^+$ is Quillen's plus construction.

This definition only holds for $n > 0$ so one often defines the higher algebraic K-theory via

$$K_n(R) = \pi_n(BGL(R)^+ \times K_0(R))$$

Since $BGL(R)^+$ is path connected and $K_0(R)$ discrete, this definition doesn't differ in higher degrees and also holds for $n = 0$.

The Q-Construction

The Q-construction gives the same results as the +-construction, but it applies in more general situations. Moreover, the definition is more direct in the sense that the K-groups, defined via the Q-construction are functorial by definition. This fact is not automatic in the plus-construction.

Suppose P is an exact category; associated to P a new category QP is defined, objects of which are those of P and morphisms from M' to M'' are isomorphism classes of diagrams

$$M' \longleftarrow N \longrightarrow M'',$$

where the first arrow is an admissible epimorphism and the second arrow is an admissible monomorphism.

The i-th K-group of the exact category P is then defined as

$$K_i(P) = \pi_{i+1}(\mathrm{B}QP, 0)$$

with a fixed zero-object 0, where BQP is the *classifying space* of QP, which is defined to be the geometric realisation of the *nerve* of QP.

This definition coincides with the above definition of $K_0(P)$. If P is the category of finitely generated projective R-modules, this definition agrees with the above BGL^+ definition of $K_n(R)$ for all n. More generally, for a scheme X, the higher K-groups of X are defined to be the K-groups of (the exact category of) locally free coherent sheaves on X.

The following variant of this is also used: instead of finitely generated projective (= locally free) modules, take finitely generated modules. The resulting K-groups are usually written $G_n(R)$. When R is a noetherian regular ring, then G- and K-theory coincide. Indeed, the global dimension of regular rings is finite, i.e. any finitely generated module has a finite projective resolution $P_* \to M$, and a simple argument shows that the canonical map $K_0(\mathrm{R}) \to G_0(\mathrm{R})$ is an isomorphism, with $[M]=\Sigma \pm [P_n]$. This isomorphism extends to the higher K-groups, too.

The S-Construction

A third construction of K-theory groups is the S-construction, due to Waldhausen. It applies to categories with cofibrations (also called Waldhausen categories). This is a more general concept than exact categories.

Examples

While the Quillen algebraic *K*-theory has provided deep insight into various aspects of algebraic geometry and topology, the *K*-groups have proved particularly difficult to compute except in a few isolated but interesting cases.

Algebraic K-Groups of Finite Fields

The first and one of the most important calculations of the higher algebraic *K*-groups of a ring were made by Quillen himself for the case of finite fields:

If $\mathbf{F}_q$ is the finite field with q elements, then:

- $K_0(\mathbf{F}_q) = \mathbf{Z}$,
- $K_{2i}(\mathbf{F}_q) = 0$ for $i \geq 1$,
- $K_{2i-1}(\mathbf{F}_q) = \mathbf{Z}/(q^i - 1)\mathbf{Z}$ for $i \geq 1$.

Algebraic K-Groups of Rings of Integers

Quillen proved that if A is the ring of algebraic integers in an algebraic number field F (a finite extension of the rationals), then the algebraic K-groups of A are finitely generated. Borel used this to calculate $K_i(A)$ and $K_i(F)$ modulo torsion. For example, for the integers **Z**, Borel proved that (modulo torsion)

- K_i (**Z**)/tors.=0 for positive i unless $i=4k+1$ with k positive
- K_{4k+1} (**Z**)/tors.= **Z** for positive k.

The torsion subgroups of $K_{2i+1}(\mathbf{Z})$, and the orders of the finite groups $K_{4k+2}(\mathbf{Z})$ have recently been determined, but whether the latter groups are cyclic, and whether the groups $K_{4k}(\mathbf{Z})$ vanish depends upon Vandiver's conjecture about the class groups of cyclotomic integers.

Applications and Open Questions

Algebraic *K*-groups are used in conjectures on special values of L-functions and the formulation of an non-commutative main conjecture of Iwasawa theory and in construction of higher regulators.

Parshin's conjecture concerns the higher algebraic *K*-groups for smooth varieties over finite fields, and states that in this case the groups vanish up to torsion.

Another fundamental conjecture due to Hyman Bass (Bass' conjecture) says that all of the groups $G_n(A)$ are finitely generated when A is a finitely generated **Z**-algebra. (The groups $G_n(A)$ are the *K*-groups of the category of finitely generated *A*-modules)

REFERENCES

1. Disquisitiones Arithmeticae at Yalepress.yale.edu
2. Elstrodt, Jürgen (2007). "The Life and Work of Gustav Lejeune Dirichlet (1805–1859)" (PDF). Clay Mathematics Proceedings. Retrieved 2007-12-25.
3. Kanemitsu, Shigeru; Chaohua Jia (2002). Number theoretic methods: future trends. Springer. pp. 271–274. ISBN 978-1-4020-1080-4.
4. Reid, Constance, 1996. Hilbert, Springer, ISBN 0-387-94674-8.
5. This work established Takagi as Japan's first mathematician of international stature.
6. Helmut Hasse, History of Class Field Theory, in Algebraic Number Theory, edited by Cassels and Frölich, Academic Press, 1967, pp. 266–279
7. Fermat's Last Theorem, Simon Singh, 1997, ISBN 1-85702-521-0>
8. Kolata, Gina (24 June 1993). "At Last, Shout of 'Eureka!' In Age-Old Math Mystery". The New York Times. Retrieved 21 January 2013.
9. See proposition VIII.8.6.11 of Neukirch, Schmidt & Wingberg 2000
10. Neukirch, Jürgen; Schmidt, Alexander; Wingberg, Kay (2000), Cohomology of Number Fields, Grundlehren der Mathematischen Wissenschaften, 323, Berlin: Springer-Verlag, ISBN 978-3-540-66671-4, Zbl 0948.11001, MR1737196
11. https://www.math.umass.edu/~weston/oldpapers/cnf.pdf
12. http://planetmath.org/realandcomplexembeddings
13. http://mathoverflow.net/questions/109330/did-gauss-know-dirichlets-class-number-formula-in-1801
14. Davenport, Harold (2000). Montgomery, Hugh L., ed. Multiplicative Number Theory. Graduate Texts in Mathematics. 74 (3rd ed.). New York: Springer-Verlag. pp. 43–53. ISBN 978-0-387-95097-6. Retrieved 2009-05-26.
15. Chevalley, Claude (1936), "Généralisation de la théorie du corps de classes pour les extensions infinies.", Journal de Mathématiques Pures et Appliquées (in French), 15: 359–371, JFM 62.1153.02
16. Chevalley, Claude (1940), "La théorie du corps de classes", Annals of Mathematics. Second Series, 41: 394–418, ISSN 0003-486X, JSTOR 1969013, MR 0002357
17. Chevalley, Claude (1951), Introduction to the Theory of Algebraic Functions of One Variable, Mathematical Surveys, No. VI, Providence, R.I.: American Mathematical Society, MR 0042164
18. Jaffard, Paul (1953), Anneaux d'adèles (d'après Iwasawa), Séminaire Bourbaki, Secrétariat mathématique, Paris, MR 0157859
19. Ono, Takashi (1957), "Sur une propriété arithmétique des groupes algébriques commutatifs", Bulletin de la Société Mathématique de France, 85: 307–323, ISSN 0037-9484, MR 0094362
20. Tate, John T. (1950), "Fourier analysis in number fields, and Hecke's zeta-functions", Algebraic Number Theory (Proc. Instructional Conf., Brighton,

1965), Thompson, Washington, D.C., pp. 305–347, ISBN 978-0-9502734-2-6, MR 0217026

21. Weil, André (1938), "Zur algebraischen Theorie der algebraischen Funktionen.", Journal für Reine und Angewandte Mathematik (in German), 179: 129–133, doi:10.1515/crll.1938.179.129, ISSN 0075-4102
22. (Neukirch 1999, Theorems VI.5.5, VI.6.1)
23. Milne, J. S. Arithmetic duality theorems. Charleston, SC: BookSurge, LLC 2006
24. Washington, Lawrence C. Introduction to cyclotomic fields. Graduate Texts in Mathematics. 83. New York, NY: Springer. 1997.
25. Artin, Emil; Tate, John (1990), Class field theory, Redwood City, Calif.: Addison-Wesley, ISBN 978-0-201-51011-9
26. Cohen, Henri; Stevenhagen, Peter (2008), "Computational class field theory", in Buhler, J.P.; P., Stevenhagen, Algorithmic Number Theory: Lattices, Number Fields, Curves and Cryptography, MSRI Publications, 44, Cambridge University Press, pp. 497–534, ISBN 978-0-521-20833-8, Zbl 1177.11095
27. Cassels, J.W.S.; Fröhlich, Albrecht, eds. (1967), Algebraic Number Theory, Academic Press, Zbl 0153.07403
28. Iwasawa, Kenkichi (1986), Local class field theory, Oxford Mathematical Monographs, The Clarendon Press Oxford University Press, ISBN 978-0-19-504030-2, MR 863740, Zbl 0604.12014
29. Neukirch, Jürgen (1986), Class Field Theory, Berlin, New York: Springer-Verlag, ISBN 978-3-540-15251-4
30. Neukirch, Jürgen (1999). Algebraic Number Theory. Grundlehren der mathematischen Wissenschaften. 322. Berlin: Springer-Verlag. ISBN 978-3-540-65399-8. Zbl 0956.11021. MR1697859.
31. Kawada, Y. (1955), "Class formations", Duke Math.J., 22: 165–177, Zbl 0067.01904
32. Kawada, Y.; Satake, I. (1956), "Class formations. II", J.Fac. Sci.Univ. Tokyo Sect. 1A, 7: 353–389, Zbl 0101.02902
33. Conrad, Keith, History of class field theory. (PDF)
34. Gras, Georges (2005), Class field theory: From theory to practice (2nd ed.), New York: Springer
35. Goldfeld, Dorian (July 1985), "Gauss' Class Number Problem For Imaginary Quadratic Fields" (PDF), Bulletin of the American Mathematical Society, 13 (1): 23–37, doi:10.1090/S0273-0979-1985-15352-2
36. Heegner, Kurt (1952), "Diophantische Analysis und Modulfunktionen", Mathematische Zeitschrift, 56 (3): 227–253, doi:10.1007/BF01174749, MR 0053135
37. te Riele, Herman; Williams, Hugh (2003), "New Computations Concerning the Cohen-Lenstra Heuristics" (PDF), Experimental Mathematics, 12 (1): 99–113, doi:10.1080/10586458.2003.10504715

38. Cohen, Henri (1993), A Course in Computational Algebraic Number Theory, Berlin: Springer, ISBN 3-540-55640-0

39. Baker, Alan (1990), Transcendental number theory, Cambridge Mathematical Library (2nd ed.), Cambridge University Press, ISBN 978-0-521-39791-9, MR 0422171

40. Bass, Hyman (1968), Algebraic K-theory, Mathematics Lecture Note Series, New York-Amsterdam: W.A. Benjamin, Inc., Zbl 0174.30302

41. Friedlander, Eric; Grayson, Daniel, eds. (2005), Handbook of K-Theory, Berlin, New York: Springer-Verlag, ISBN 978-3-540-30436-4, MR 2182598

42. Friedlander, Eric M.; Weibel, Charles W. (1999), An overview of algebraic K-theory, World Sci. Publ., River Edge, NJ, pp. 1–119, MR 1715873

43. Gille, Philippe; Szamuely, Tamás (2006), Central simple algebras and Galois cohomology, Cambridge Studies in Advanced Mathematics, 101, Cambridge: Cambridge University Press, ISBN 0-521-86103-9, Zbl 1137.12001

44. Gras, Georges (2003), Class field theory. From theory to practice, Springer Monographs in Mathematics, Berlin: Springer-Verlag, ISBN 3-540-44133-6, Zbl 1019.11032

45. Lam, Tsit-Yuen (2005), Introduction to Quadratic Forms over Fields, Graduate Studies in Mathematics, 67, American Mathematical Society, ISBN 0-8218-1095-2, MR 2104929, Zbl 1068.11023

46. Lemmermeyer, Franz (2000), Reciprocity laws. From Euler to Eisenstein, Springer Monographs in Mathematics, Berlin: Springer-Verlag, doi:10.1007/978-3-662-12893-0, ISBN 3-540-66957-4, MR 1761696, Zbl 0949.11002

47. Milnor, John Willard (1970), "Algebraic K-theory and quadratic forms", Inventiones Mathematicae, 9 (4): 318–344, doi:10.1007/BF01425486, ISSN 0020-9910, MR 0260844

48. Milnor, John Willard (1971), Introduction to algebraic K-theory, Annals of Mathematics Studies, 72, Princeton, NJ: Princeton University Press, MR 0349811, Zbl 0237.18005 (lower K-groups)

49. Quillen, Daniel (1973), "Higher algebraic K-theory. I", Algebraic K-theory, I: Higher K-theories (Proc. Conf., Battelle Memorial Inst., Seattle, Wash., 1972), Lecture Notes in Math, 341, Berlin, New York: Springer-Verlag, pp. 85–147, doi:10.1007/BFb0067053, ISBN 978-3-540-06434-3, MR 0338129

50. Quillen, Daniel (1975), "Higher algebraic K-theory", Proceedings of the International Congress of Mathematicians (Vancouver, B. C., 1974), Vol. 1, Montreal, Quebec: Canad. Math. Congress, pp. 171–176, MR 0422392 (Quillen's Q-construction)

51. Quillen, Daniel (1974), "Higher K-theory for categories with exact sequences", New developments in topology (Proc. Sympos. Algebraic Topology, Oxford, 1972), London Math. Soc. Lecture Note Ser., 11, Cambridge University Press, pp. 95–103, MR 0335604 (relation of Q-construction to plus-construction)

52. Rosenberg, Jonathan (1994), Algebraic K-theory and its applications, Graduate

Texts in Mathematics, 147, Berlin, New York: Springer-Verlag, ISBN 978-0-387-94248-3, MR 1282290, Zbl 0801.19001. Errata

53. Seiler, Wolfgang (1988), "λ-Rings and Adams Operations in Algebraic K-Theory", in Rapoport, M.; Schneider, P.; Schappacher, N., Beilinson's Conjectures on Special Values of L-Functions, Boston, MA: Academic Press, ISBN 978-0-12-581120-0
54. Silvester, John R. (1981), Introduction to algebraic K-theory, Chapman and Hall Mathematics Series, London, New York: Chapman and Hall, ISBN 0-412-22700-2, Zbl 0468.18006
55. Weibel, Charles (2005), "Algebraic K-theory of rings of integers in local and global fields", Handbook of K-theory (PDF), Berlin, New York: Springer-Verlag, pp. 139–190, MR 2181823 (survey article)
56. Weibel, C., The Development of Algebraic K-theory before 1980, AMS Contemp. Math. 243 (1999), 211-238.

CHAPTER 4

Analytic Number Theory

Chapter Contents

Analytic number theory is the branch of number theory which uses real and complex analysis to investigate various properties of integers and prime numbers. Examples of topics falling under analytic number theory include Dirichlet L-series, the Riemann zeta function $\zeta(s)$, the totient function $\phi(n)$, and the prime number theorem.

INTRODUCTION

In mathematics, analytic number theory is a branch of number theory that uses methods from mathematical analysis to solve problems about the integers. It is often said to have begun with Peter Gustav Lejeune Dirichlet's 1837 introduction of Dirichlet L-functions to give the first proof of Dirichlet's theorem on arithmetic progressions. It is well known for its results on prime numbers (involving the Prime Number Theorem and Riemann zeta function) and additive number theory (such as the Goldbach conjecture and Waring's problem).

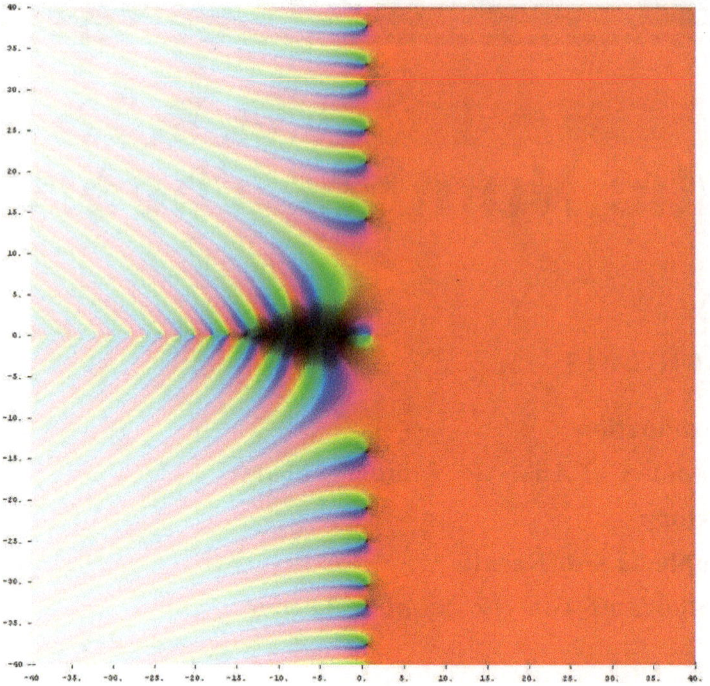

Riemann zeta function (s) in the complex plane. The color of a point s encodes the value of (s): colors close to black denote values close to zero, while hue encodes the value's argument.

BRANCHES OF ANALYTIC NUMBER THEORY

Analytic number theory can be split up into two major parts, divided more by the type of problems they attempt to solve than fundamental differences in technique.

- Multiplicative number theory deals with the distribution of the prime numbers, such as estimating the number of primes in an interval, and includes the prime number theorem and Dirichlet's theorem on primes in arithmetic progressions.
- Additive number theory is concerned with the additive structure of the integers, such as Goldbach's conjecture that every even number greater than 2 is the sum of two primes. One of the main results in additive number theory is the solution to Waring's problem.

HISTORY

Precursors

Much of analytic number theory was inspired by the prime number theorem. Let $\pi(x)$ be the prime-counting function that gives the number of primes less than or equal to x, for any real number x. For example, $\pi(10) = 4$ because there are four prime numbers

(2, 3, 5 and 7) less than or equal to 10. The prime number theorem then states that x / ln(x) is a good approximation to π(x), in the sense that the limit of the quotient of the two functions π(x) and x / ln(x) as x approaches infinity is 1:

$$\lim_{x\to\infty} \frac{\pi(x)}{x/\ln(x)} = 1,$$

known as the asymptotic law of distribution of prime numbers.

Adrien-Marie Legendre conjectured in 1797 or 1798 that π(a) is approximated by the function a/(A ln(a) + B), where A and B are unspecified constants. In the second edition of his book on number theory (1808) he then made a more precise conjecture, with A = 1 and B = −1.08366. Carl Friedrich Gauss considered the same question: "Im Jahr 1792 oder 1793", according to his own recollection nearly sixty years later in a letter to Encke (1849), he wrote in his logarithm table (he was then 15 or 16) the short note "Primzahlen unter $a(=\infty)\frac{a}{\ln a}$". But Gauss never published this conjecture. In 1838 Peter Gustav Lejeune Dirichlet came up with his own approximating function, the logarithmic integral li(x) (under the slightly different form of a series, which he communicated to Gauss). Both Legendre's and Dirichlet's formulas imply the same conjectured asymptotic equivalence of π(x) and x / ln(x) stated above, although it turned out that Dirichlet's approximation is considerably better if one considers the differences instead of quotients.

Dirichlet

Johann Peter Gustav Lejeune Dirichlet is credited with the creation of analytic number theory, a field in which he found several deep results and in proving them introduced some fundamental tools, many of which were later named after him. In 1837 he published Dirichlet's theorem on arithmetic progressions, using mathematical analysis concepts to tackle an algebraic problem and thus creating the branch of analytic number theory. In proving the theorem, he introduced the Dirichlet characters and L-functions. In 1841 he generalized his arithmetic progressions theorem from integers to the ring of Gaussian integers $\mathbb{Z}[i]$.

Chebyshev

In two papers from 1848 and 1850, the Russian mathematician Pafnuty L'vovich Chebyshev attempted to prove the asymptotic law of distribution of prime numbers. His work is notable for the use of the zeta function ζ(s) (for real values of the argument "s", as are works of Leonhard Euler, as early as 1737) predating Riemann's celebrated memoir of 1859, and he succeeded in proving a slightly weaker form of the asymptotic law, namely, that if the limit of π(x)/(x/ln(x)) as x goes to infinity exists at all, then it is necessarily equal to one. He was able to prove unconditionally that this ratio is bounded above and below by two explicitly given constants near to 1 for all x. Although Chebyshev's paper did not prove the Prime Number Theorem, his estimates for π(x) were strong enough for him to prove Bertrand's postulate that there exists a prime number between n and 2n for any integer $n \geq 2$.

Riemann

"...es ist sehr wahrscheinlich, dass alle Wurzeln reell sind. Hiervon wäre allerdings ein strenger Beweis zu wünschen; ich habe indess die Aufsuchung desselben nach einigen flüchtigen vergeblichen Versuchen vorläufig bei Seite gelassen, da er für den nächsten Zweck meiner Untersuchung entbehrlich schien."

„...it is very probable that all roots are real. Of course one would wish for a rigorous proof here; I have for the time being, after some fleeting vain attempts, provisionally put aside the search for this, as it appears dispensable for the next objective of my investigation."

Riemann's statement of the Riemann hypothesis, from his 1859 paper. (He was discussing a version of the zeta function, modified so that its roots are real rather than on the critical line.)

Bernhard Riemann made some famous contributions to modern analytic number theory. In a single short paper (the only one he published on the subject of number theory), he investigated the Riemann zeta function and established its importance for understanding the distribution of prime numbers. He made a series of conjectures about properties of the zeta function, one of which is the well-known Riemann hypothesis.

Hadamard and de la Vallée-Poussin

Extending the ideas of Riemann, two proofs of the prime number theorem were obtained independently by Jacques Hadamard and Charles Jean de la Vallée-Poussin and appeared in the same year (1896). Both proofs used methods from complex analysis, establishing as a main step of the proof that the Riemann zeta function $\zeta(s)$ is non-zero for all complex values of the variable s that have the form $s = 1 + it$ with $t > 0$.

Modern Times

The biggest technical change after 1950 has been the development of sieve methods, particularly in multiplicative problems. These are combinatorial in nature, and quite varied. The extremal branch of combinatorial theory has in return been greatly influenced by the value placed in analytic number theory on quantitative upper and lower bounds. Another recent development is probabilistic number theory, which uses methods from probability theory to estimate the distribution of number theoretic functions, such as how many prime divisors a number has.

Developments within analytic number theory are often refinements of earlier techniques, which reduce the error terms and widen their applicability. For example, the circle method of Hardy and Littlewood was conceived as applying to power series near the unit circle in the complex plane; it is now thought of in terms of finite exponential sums (that is, on the unit circle, but with the power series truncated). The needs of diophantine approximation are for auxiliary functions that are not generating functions—their coefficients are constructed by use of a pigeonhole principle—and involve several complex variables. The fields of diophantine approximation and transcendence theory have expanded, to the point that the techniques have been applied to the Mordell conjecture.

PROBLEMS AND RESULTS

Theorems and results within analytic number theory tend not to be exact structural results about the integers, for which algebraic and geometrical tools are more appropriate. Instead, they give approximate bounds and estimates for various number theoretical functions, as the following examples illustrate.

Multiplicative Number Theory

Multiplicative number theory is a subfield of analytic number theory that deals with prime numbers and with factorization and divisors. The focus is usually on developing approximate formulas for counting these objects in various contexts. The prime number theorem is a key result in this subject. The Mathematics Subject Classification for multiplicative number theory is 11Nxx.

Scope

Multiplicative number theory deals primarily in asymptotic estimates for arithmetic functions. Historically the subject has been dominated by the prime number theorem, first by attempts to prove it and then by improvements in the error term. The Dirichlet divisor problem that estimates the average order of the divisor function d(n) and Gauss's circle problem that estimates the average order of the number of representations of a number as a sum of two squares are also classical problems, and again the focus is on improving the error estimates.

The distribution of primes numbers among residue classes modulo an integer is an area of active research. Dirichlet's theorem on primes in arithmetic progressions shows that there are an infinity of primes in each co-prime residue class, and the prime number theorem for arithmetic progressions shows that the primes are asymptotically equidistributed among the residue classes. The Bombieri–Vinogradov theorem gives a more precise measure of how evenly they are distributed. There is also much interest in the size of the smallest prime in an arithmetic progression; Linnik's theorem gives an estimate.

The twin prime conjecture, namely that there are an infinity of primes p such that p+2 is also prime, is the subject of active research. Chen's theorem shows that there are an infinity of primes p such that p+2 is either prime or the product of two primes.

Methods

The methods belong primarily to analytic number theory, but elementary methods, especially sieve methods, are also very important. The large sieve and exponential sums are usually considered part of multiplicative number theory.

The distribution of prime numbers is closely tied to the behavior of the Riemann zeta function and the Riemann hypothesis, and these subjects are studied both from a number theory viewpoint and a complex analysis viewpoint.

Euclid showed that there are an infinite number of primes but it is very difficult to find an efficient method for determining whether or not a number is prime, especially a large number. A related but easier problem is to determine the asymptotic distribution of the prime numbers; that is, a rough description of how many primes are smaller than a given number. Gauss, amongst others, after computing a large list of primes, conjectured that the number of primes less than or equal to a large number N is close to the value of the integral

$$\int_2^N \frac{1}{\log t}\,dt.$$

In 1859 Bernhard Riemann used complex analysis and a special meromorphic function now known as the Riemann zeta function to derive an analytic expression for the number of primes less than or equal to a real number x. Remarkably, the main term in Riemann's formula was exactly the above integral, lending substantial weight to Gauss's conjecture. Riemann found that the error terms in this expression, and hence the manner in which the primes are distributed, are closely related to the complex zeros of the zeta function. Using Riemann's ideas and by getting more information on the zeros of the zeta function, Jacques Hadamard and Charles Jean de la Vallée-Poussin managed to complete the proof of Gauss's conjecture. In particular, they proved that if

$$\pi(x) = (\textbf{number of primes } \leq x),$$

then

$$\lim_{x\to\infty} \frac{\pi(x)}{x/\log x} = 1.$$

This remarkable result is what is now known as the Prime Number Theorem. It is a central result in analytic number theory. Loosely speaking, it states that given a large number N, the number of primes less than or equal to N is about N/log(N).

More generally, the same question can be asked about the number of primes in any arithmetic progression a+nq for any integer n. In one of the first applications of analytic techniques to number theory, Dirichlet proved that any arithmetic progression with a and q coprime contains infinitely many primes. The prime number theorem can be generalised to this problem; letting

$$\pi(x,a,q) = (\textbf{number of primes } \leq x \textbf{ such that } p \textbf{ is in the arithmetic progression } a+nq, n \in \mathbf{Z}),$$

then if a and q are coprime,

$$\lim_{x\to\infty} \frac{\pi(x,a,q)\phi(q)}{x/\log x} = 1.$$

There are also many deep and wide ranging conjectures in number theory whose proofs seem too difficult for current techniques, such as the Twin prime conjecture which asks whether there are infinitely many primes p such that p + 2 is prime. On the assumption of the Elliott–Halberstam conjecture it has been proven recently that there are infinitely many primes p such that p + k is prime for some positive even k at most 12. Also, it has been proven unconditionally (i.e. not depending on unproven conjectures) that there are infinitely many primes p such that p + k is prime for some positive even k at most 246.

Additive Number Theory

In number theory, the specialty additive number theory studies subsets of integers and their behavior under addition. More abstractly, the field of "additive number theory" includes the study of abelian groups and commutative semigroups with an operation of addition. Additive number theory has close ties to combinatorial number theory and the geometry of numbers. Two principal objects of study are the sumset of two subsets A and B of elements from an abelian group G,

$$A + B = \{a + b : a \in A, b \in B\},$$

and the h-fold sumset of A,

$$hA = \underbrace{A + \cdots + A}_{h}.$$

There are two main subdivisions listed below.

The first is principally devoted to consideration of direct problems over (typically) the integers, that is, determining the structure of hA from the structure of A: for example, determining which elements can be represented as a sum from hA, where A is a fixed subset. Two classical problems of this type are the Goldbach conjecture (which is the conjecture that 2P contains all even numbers greater than two, where P is the set of primes) and Waring's problem (which asks how large must h be to guarantee that hA_k contains all positive integers, where

$$A_k = \{0^k, 1^k, 2^k, 3^k, \ldots\}$$

is the set of k-th powers). Many of these problems are studied using the tools from the Hardy-Littlewood circle method and from sieve methods. For example, Vinogradov proved that every sufficiently large odd number is the sum of three primes, and so every sufficiently large even integer is the sum of four primes. Hilbert proved that, for every integer k > 1, every nonnegative integer is the sum of a bounded number of k-th powers. In general, a set A of nonnegative integers is called a basis of order h if hA contains all positive integers, and it is called an asymptotic basis if hA contains all sufficiently large integers. Much current research in this area concerns properties of general asymptotic bases of finite order. For example, a set A is called a minimal asymptotic basis of order h if A is an asymptotic basis of order h but no proper subset of A is an asymptotic basis of order h. It has been proved that minimal asymptotic bases of order h exist for all h, and that there also exist asymptotic bases of order h that contain no minimal asymptotic bases of order h. Another question to be considered is how small can the number of representations of n as a sum of h elements in an asymptotic basis can be. This is the content of the Erdős–Turán conjecture on additive bases.

Additive Combinatorics

The second is principally devoted to consideration of inverse problems, often over more general groups than just the integers, that is, given some information about the sumset A+B, the aim is find information about the structure of the individual sets A and B. (A more recent name sometimes associated to this sub-division is additive combinatorics.) Unlike problems related to classical bases, as described above, this sub-area often deals with finite subsets rather than infinite ones. A typical question is what is the structure of a pair of subsets whose sumset has small cardinality (in relation to |A| and |B|). In the case of the integers, the classical Freiman's theorem provides a potent partial answer to this question in terms of multi-dimensional arithmetic progressions. Another typical problem is simply to find a lower bound for |A+B| in terms of |A| and |B| (this can be view as an inverse problem with the given information for A+B being that |A+B| is sufficiently small and the structural conclusion then being that either A or B is the empty set; such problems are often considered direct problems as well). Examples of this type include the Erdős–Heilbronn Conjecture (for a restricted sumset) and the Cauchy–Davenport Theorem. The methods used for tackling such questions draw from across the spectrum of mathematics, including combinatorics, ergodic theory, analysis, graph theory, group theory, and linear algebraic and polynomial methods.

Diophantine Problems

In mathematics, a Diophantine equation is a polynomial equation, usually in two or more unknowns, such that only the integer solutions are sought or studied (an integer solution is a solution such that all the unknowns take integer values). A linear Diophantine equation is an equation between two sums of monomials of degree zero or one. An exponential Diophantine equation is one in which exponents on terms can be unknowns.

Diophantine problems have fewer equations than unknown variables and involve finding integers that work correctly for all equations. In more technical language, they define an algebraic curve, algebraic surface, or more general object, and ask about the lattice points on it.

The word Diophantine refers to the Hellenistic mathematician of the 3rd century, Diophantus of Alexandria, who made a study of such equations and was one of the first mathematicians to introduce symbolism into algebra. The mathematical study of Diophantine problems that Diophantus initiated is now called Diophantine analysis.

While individual equations present a kind of puzzle and have been considered throughout history, the formulation of general theories of Diophantine equations (beyond the theory of quadratic forms) was an achievement of the twentieth century.

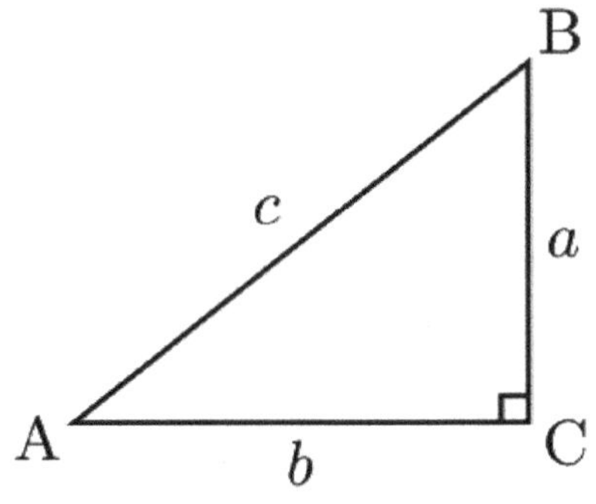

Finding all right triangles with integer side-lengths is equivalent to solving the Diophantine equation $a^2 + b^2 = c^2$.

Diophantine problems are concerned with integer solutions to polynomial equations: one may study the distribution of solutions, that is, counting solutions according to some measure of "size" or height.

An important example is the Gauss circle problem, which asks for integers points (x y) which satisfy

$$x^2 + y^2 \leq r^2.$$

In geometrical terms, given a circle centered about the origin in the plane with radius r, the problem asks how many integer lattice points lie on or inside the circle. It is not hard to prove that the answer is $\pi r^2 + E(r)$, where $E(r)/r^2 \to 0$ as $r \to \infty$. Again, the difficult part and a great achievement of analytic number theory is obtaining specific upper bounds on the error term E(r).

It was shown by Gauss that $E(r) = O(r)$. In general, an O(r) error term would be possible with the unit circle (or, more properly, the closed unit disk) replaced by the dilates of any bounded planar region with piecewise smooth boundary. Furthermore, replacing the unit circle by the unit square, the error term for the general problem can be as large as a linear function of r. Therefore, getting an error bound of the form $O(r^\delta)$ for some $\delta < 1$ in the case of the circle is a significant improvement. The first to attain this was Sierpiński in 1906, who showed $E(r) = O(r^{2/3})$. In 1915, Hardy and Landau each showed that one does not have $E(r) = O(r^{1/2})$. Since then the goal has been to show that for each fixed $\epsilon > 0$ there exists a real number $C(\epsilon)$ such that $E(r) \leq C(\epsilon) r^{1/2+\epsilon}$.

In 2000 Huxley showed that $E(r) = O(r^{131/208})$, which is the best published result.

Examples

In the following Diophantine equations, w, x, y, and z are the unknowns and the other letters are given constants:

$ax + by = 1$	This is a linear Diophantine equation.
$w^3 + x^3 = y^3 + z^3$	The smallest nontrivial solution in positive integers is $12^3 + 1^3 = 9^3 + 10^3 = 1729$. It was famously given as an evident property of 1729, a taxicab number (also named Hardy–Ramanujan number) by Ramanujan to Hardy while meeting in 1917. There are infinitely many nontrivial solutions.
$x^n + y^n = z^n$	For n = 2 there are infinitely many solutions (x,y,z): the Pythagorean triples. For larger integer values of n, Fermat's Last Theorem (initially claimed in 1637 by Fermat and proved by Wiles in 1995) states there are no positive integer solutions (x,y,z).
$x^2 - ny^2 = \pm 1$	This is Pell's equation, which is named after the English mathematician John Pell. It was studied by Brahmagupta in the 7th century, as well as by Fermat in the 17th century.
$4/n = 1/x + 1/y + 1/z$	The Erdős–Straus conjecture states that, for every positive integer $n \geq 2$, there exists a solution in x, y, and z, all as positive integers. Although not usually stated in polynomial form, this example is equivalent to the polynomial equation $4xyz = yzn + xzn + xyn = n(yz + xz + xy)$.
$x^4 + y^4 + z^4 = w^4$	Conjectured incorrectly by Euler to have no nontrivial solutions. Proved by Elkies to have infinitely many nontrivial solutions, with a computer search by Frye determining the smallest nontrivial solution.

Linear Diophantine Equations

One Equation

The simplest linear Diophantine equation takes the form $ax + by = c$, where a, b and c are given integers. The solutions are described by the following theorem:

This Diophantine equation has a solution (where x and y are integers) if and only if c is a multiple of the greatest common divisor of a and b. Moreover, if (x, y) is a solution, then the other solutions have the form $(x + kv, y - ku)$, where k is an arbitrary integer, and u and v are the quotients of a and b (respectively) by the greatest common divisor of a and b.

Proof: If d is this greatest common divisor, Bézout's identity asserts the existence of integers e and f such that $ae + bf = d$. If c is a multiple of d, then $c = dh$ for some integer h, and (eh, fh) is a solution. On the other hand, for every pair of integers x and y, the greatest common divisor d of a and b divides $ax + by$. Thus, if the equation has a solution, then c must be a multiple of d. If $a = ud$ and $b = vd$, then for every solution (x, y), we have

$$a(x + kv) + b(y - ku) = ax + by + k(av - bu) = ax + by + k(udv - vdu) = ax + by,$$

showing that $(x + kv, y - ku)$ is another solution. Finally, given two solutions such that $ax_1 + by_1 = ax_2 + by_2 = c$, one deduces that $u(x_2 - x_1) + v(y_2 - y_1) = 0$. As u and v are coprime, Euclid's lemma shows that there exists an integer k such that $x_2 - x_1 = kv$ and $y_2 - y_1 = -ku$. Therefore, $x_2 = x_1 + kv$ and $y_2 = y_1 - ku$, which completes the proof.

Chinese Remainder Theorem

The Chinese remainder theorem describes an important class of linear Diophantine systems of equations: let $n_1, \ldots, n_k$ be k pairwise coprime integers greater than one, $a_1, \ldots, a_k$ be karbitrary integers, and N be the product $n_1 \cdots n_k$. The Chinese remainder theorem asserts that the following linear Diophantine system has exactly one solution $(x, x_1, \ldots, x_k)$ such that $0 \leq x < N$, and that the other solutions are obtained by adding to x a multiple of N:

$$x = a_1 + n_1\,x_1$$

$$\vdots$$

$$x = a_k + n_k\,x_k$$

System of Linear Diophantine Equations

More generally, every system of linear Diophantine equations may be solved by computing the Smith normal form of its matrix, in a way that is similar to the use of the reduced row echelon form to solve a system of linear equations over a field. Using matrix notation every system of linear Diophantine equations may be written

$AX = C$,

where A is an $m \times n$ matrix of integers, X is an $n \times 1$ column matrix of unknowns and C is an $m \times 1$ column matrix of integers.

The computation of the Smith normal form of A provides two unimodular matrices (that is matrices that are invertible over the integers and have ±1 as determinant) U and V of respective dimensions $m \times m$ and $n \times n$, such that the matrix

$B = [b_{i,j}] = UAV$

is such that $b_{i,i}$ is not zero for i not greater than some integer k, and all the other entries are zero. The system to be solved may thus be rewritten as

$B(V^{-1}X) = UC$.

Calling y_i the entries of $V^{-1}X$ and d_i those of $D = UC$, this leads to the system

$b_{i,i}y_i = d_i$ for $1 \leq i \leq k$,

$0y_i = d_i$ for $k < i \leq n$.

This system is equivalent to the given one in the following sense: A column matrix of integers x is a solution of the given system if and only if $x = Vy$ for some column matrix of integers ysuch that $By = D$.

It follows that the system has a solution if and only if $b_{i,i}$ divides d_i for $i \leq k$ and $d_i = 0$ for $i > k$. If this condition is fulfilled, the solutions of the given system are

$$V\begin{bmatrix} \frac{d_1}{b_{1,1}} \\ \vdots \\ \frac{d_k}{b_{k,k}} \\ h_{k+1} \\ \vdots \\ h_n \end{bmatrix},$$

where $h_{k+1}, \ldots, h_n$ are arbitrary integers.

Hermite normal form may also be used for solving systems of linear Diophantine equations. However, Hermite normal form does not directly provide the solutions; to get the solutions from the Hermite normal form, one has to successively solve several linear equations. Nevertheless, Richard Zippel wrote that the Smith normal form "is somewhat more than is actually needed to solve linear diophantine equations. Instead of reducing the equation to diagonal form, we only need to make it triangular, which is called the Hermite normal form. The Hermite normal form is substantially easier to compute than the Smith normal form."

Integer linear programming amounts to finding some integer solutions (optimal in some sense) of linear systems that include also inequations. Thus systems of linear Diophantine equations are basic in this context, and textbooks on integer programming usually have a treatment of systems of linear Diophantine equations.

Diophantine Analysis

Typical Questions

The questions asked in Diophantine analysis include:

- Are there any solutions?
- Are there any solutions beyond some that are easily found by inspection?
- Are there finitely or infinitely many solutions?
- Can all solutions be found in theory?
- Can one in practice compute a full list of solutions?

These traditional problems often lay unsolved for centuries, and mathematicians gradually came to understand their depth (in some cases), rather than treat them as puzzles.

Typical Problem

The given information is that a father's age is 1 less than twice that of his son, and that the digits AB making up the father's age are reversed in the son's age (i.e. BA). This leads to the equation $10A + B = 2(10B + A) - 1$, thus $19B - 8A = 1$. Inspection gives the result $A = 7$, $B = 3$, and thus AB equals 73 years and BA equals 37 years. One may easily show that there is not any other solution with A and B positive integers less than 10.

17th and 18th Centuries

In 1637, Pierre de Fermat scribbled on the margin of his copy of Arithmetica: "It is impossible to separate a cube into two cubes, or a fourth power into two fourth powers, or in general, any power higher than the second into two like powers." Stated in more modern language, "The equation $a^n + b^n = c^n$ has no solutions for any n higher than 2." And then he wrote, intriguingly: "I have discovered a truly marvelous proof of this proposition, which this margin is too narrow to contain." Such a proof eluded mathematicians for centuries, however, and as such his statement became famous as Fermat's Last Theorem. It wasn't until 1995 that it was proven by the British mathematician Andrew Wiles.

In 1657, Fermat attempted to solve the Diophantine equation $61x^2 + 1 = y^2$ (solved by Brahmagupta over 1000 years earlier). The equation was eventually solved by Euler in the early 18th century, who also solved a number of other Diophantine equations. The smallest solution of this equation in positive integers is $x = 226153980$, $y = 1766319049$.

Hilbert's Tenth Problem

In 1900, David Hilbert proposed the solvability of all Diophantine equations as the tenth of his fundamental problems. In 1970, Yuri Matiyasevich solved it negatively, by proving that a general algorithm for solving all Diophantine equations cannot exist.

Diophantine Geometry

Diophantine geometry, which is the application of techniques from algebraic geometry in this field, has continued to grow as a result; since treating arbitrary equations is a dead end, attention turns to equations that also have a geometric meaning. The central idea of Diophantine geometry is that of a rational point, namely a solution to a polynomial equation or a system of polynomial equations, which is a vector in a prescribed field K, when K is not algebraically closed.

Modern Research

One of the few general approaches is through the Hasse principle. Infinite descent is the traditional method, and has been pushed a long way.

The depth of the study of general Diophantine equations is shown by the characterization of Diophantine sets as equivalently described as recursively enumerable. In other words, the general problem of Diophantine analysis is blessed or cursed with universality, and in any case is not something that will be solved except by re-expressing it in other terms.

The field of Diophantine approximation deals with the cases of Diophantine inequalities. Here variables are still supposed to be integral, but some coefficients may be irrational numbers, and the equality sign is replaced by upper and lower bounds.

The most celebrated single question in the field, the conjecture known as Fermat's Last Theorem, was solved by Andrew Wiles but using tools from algebraic geometry developed during the last century rather than within number theory where the conjecture was originally formulated. Other major results, such as Faltings' theorem, have disposed of old conjectures.

Infinite Diophantine Equations

An example of an infinite diophantine equation is:

$n = a^2 + 2b^2 + 3c^2 + 4d^2 + 5e^2 + \ldots,$

which can be expressed as "How many ways can a given integer n be written as the sum of a square plus twice a square plus thrice a square and so on?" The number of ways this can be done for each n forms an integer sequence. Infinite Diophantine equations are related to theta functions and infinite dimensional lattices. This equation always has a solution for any positive n. Compare this to:

$n = a^2 + 4b^2 + 9c^2 + 16d^2 + 25e^2 + \ldots,$

which does not always have a solution for positive n.

Exponential Diophantine Equations

If a Diophantine equation has as an additional variable or variables occurring as exponents, it is an exponential Diophantine equation. Examples include the Ramanujan–Nagell equation, $2^n - 7 = x^2$, and the equation of the Fermat-Catalan conjecture and Beal's conjecture, $a^m + b^n = c^k$ with inequality restrictions on the exponents. A general theory for such equations is not available; particular cases such as Catalan's conjecture have been tackled. However, the majority are solved via ad hoc methods such as Størmer's theorem or even trial and error.

METHODS OF ANALYTIC NUMBER THEORY

Dirichlet Series

In mathematics, a Dirichlet series is any series of the form

$$\sum_{n=1}^{\infty} \frac{a_n}{n^s},$$

where s is complex, and a is a complex sequence. It is a special case of general Dirichlet series.

Dirichlet series play a variety of important roles in analytic number theory. The most usually seen definition of the Riemann zeta function is a Dirichlet series, as are the Dirichlet L-functions. It is conjectured that the Selberg class of series obeys the generalized Riemann hypothesis. The series is named in honor of Peter Gustav Lejeune Dirichlet.

Combinatorial Importance

Dirichlet series can be used as generating series for counting weighted sets of objects with respect to a weight which is combined multiplicatively when taking Cartesian products.

Suppose that A is a set with a function w: A → **N** assigning a weight to each of the elements of A, and suppose additionally that the fibre over any natural number under that weight is a finite set. (We call such an arrangement (A,w) a weighted set.) Suppose additionally that a_n is the number of elements of A with weight n. Then we define the formal Dirichlet generating series for A with respect to w as follows:

$$\mathfrak{D}_w^A(s) = \sum_{a \in A} \frac{1}{w(a)^s} = \sum_{n=1}^{\infty} \frac{a_n}{n^s}$$

Note that if A and B are disjoint subsets of some weighted set (U, w), then the Dirichlet series for their (disjoint) union is equal to the sum of their Dirichlet series:

$$\mathfrak{D}_w^{A \uplus B}(s) = \mathfrak{D}_w^A(s) + \mathfrak{D}_w^B(s).$$

Moreover, and perhaps a bit more interestingly, if (A, u) and (B, v) are two weighted sets, and we define a weight function w: A × B → **N** by

$$w(a, b) = u(a)v(b),$$

for all a in A and b in B, then we have the following decomposition for the Dirichlet series of the Cartesian product:

$$\mathfrak{D}_w^{A \times B}(s) = \mathfrak{D}_u^A(s) \cdot \mathfrak{D}_v^B(s).$$

This follows ultimately from the simple fact that $n^{-s} \cdot m^{-s} = (nm)^{-s}$

Examples

The most famous of Dirichlet series is

$$\zeta(s) = \sum_{n=1}^{\infty} \frac{1}{n^s},$$

which is the Riemann zeta function.

Treating these as formal Dirichlet series for the time being in order to be able to ignore matters of convergence, note that we have:

$$\zeta(s) = \mathfrak{D}_{\mathrm{id}}^{\mathrm{N}}(s) = \prod_{p \text{ prime}} \mathfrak{D}_{\mathrm{id}}^{\{p^n : n \in \mathrm{N}\}}(s) = \prod_{p \text{ prime}} \sum_{n \in \mathrm{N}} \mathfrak{D}_{\mathrm{id}}^{\{p^n\}}(s)$$

$$= \prod_{p \text{ prime}} \sum_{n \in \mathrm{N}} \frac{1}{(p^n)^s} = \prod_{p \text{ prime}} \sum_{n \in \mathrm{N}} \left(\frac{1}{p^s}\right)^n = \prod_{p \text{ prime}} \frac{1}{1-p^{-s}},$$

as each natural number has a unique multiplicative decomposition into powers of primes. It is this bit of combinatorics which inspires the Euler product formula.

Another is:

$$\frac{1}{\zeta(s)} = \sum_{n=1}^{\infty} \frac{\mu(n)}{n^s}$$

where $\mu(n)$ is the Möbius function. This and many of the following series may be obtained by applying Möbius inversion and Dirichlet convolution to known series. For example, given a Dirichlet character $\chi(n)$ one has

$$\frac{1}{L(\chi, s)} = \sum_{n=1}^{\infty} \frac{\mu(n)\chi(n)}{n^s}$$

where $L(\chi, s)$ is a Dirichlet L-function.

Other identities include

$$\frac{\zeta(s-1)}{\zeta(s)} = \sum_{n=1}^{\infty} \frac{\varphi(n)}{n^s}$$

where $\varphi(n)$ is the totient function,

$$\frac{\zeta(s-k)}{\zeta(s)} = \sum_{n=1}^{\infty} \frac{J_k(n)}{n^s}$$

where J_k is the Jordan function, and

$$\zeta(s)\zeta(s-a) = \sum_{n=1}^{\infty} \frac{\sigma_a(n)}{n^s}$$

$$\frac{\zeta(s)\zeta(s-a)\zeta(s-2a)}{\zeta(2s-2a)} = \sum_{n=1}^{\infty} \frac{\sigma_a(n^2)}{n^s}$$

$$\frac{\zeta(s)\zeta(s-a)\zeta(s-b)\zeta(s-a-b)}{\zeta(2s-a-b)} = \sum_{n=1}^{\infty} \frac{\sigma_a(n)\sigma_b(n)}{n^s}$$

where $_a(n)$ is the divisor function. By specialization to the divisor function $d=\sigma_0$ we have

$$\zeta^2(s) = \sum_{n=1}^{\infty} \frac{d(n)}{n^s}$$

$$\frac{\zeta^3(s)}{\zeta(2s)} = \sum_{n=1}^{\infty} \frac{d(n^2)}{n^s}$$

$$\frac{\zeta^4(s)}{\zeta(2s)} = \sum_{n=1}^{\infty} \frac{d(n)^2}{n^s}.$$

The logarithm of the zeta function is given by

$$\log \zeta(s) = \sum_{n=2}^{\infty} \frac{\Lambda(n)}{\log(n)} \frac{1}{n^s}$$

for Re(s) > 1. Here, $\Lambda(n)$ is the von Mangoldt function. The logarithmic derivative is then

$$\frac{\zeta'(s)}{\zeta(s)} = -\sum_{n=1}^{\infty} \frac{\Lambda(n)}{n^s}.$$

These last two are special cases of a more general relationship for derivatives of Dirichlet series, given below.

Given the Liouville function $\lambda(n)$, one has

$$\frac{\zeta(2s)}{\zeta(s)} = \sum_{n=1}^{\infty} \frac{\lambda(n)}{n^s}.$$

Yet another example involves Ramanujan's sum:

$$\frac{\sigma_{1-s}(m)}{\zeta(s)} = \sum_{n=1}^{\infty} \frac{c_n(m)}{n^s}.$$

Another example involves the Möbius function:

$$\frac{\zeta(s)}{\zeta(2s)} = \sum_{n=1}^{\infty} \frac{|\mu(n)|}{n^s} \equiv \sum_{n=1}^{\infty} \frac{\mu^2(n)}{n^s}.$$

Analytic properties of Dirichlet series

Given a sequence $\{a_n\}_{n \in N}$ of complex numbers we try to consider the value of

$$f(s) = \sum_{n=1}^{\infty} \frac{a_n}{n^s}$$

as a function of the complex variable s. In order for this to make sense, we need to consider the convergence properties of the above infinite series:

If $\{a_n\}_{n \in N}$ is a bounded sequence of complex numbers, then the corresponding Dirichlet series f converges absolutely on the open half-plane of s such that Re(s) > 1. In general, if $a_n = O(n^k)$, the series converges absolutely in the half plane Re(s) > k + 1.

If the set of sums $a_n + a_{n+1} + ... + a_{n+k}$ is bounded for n and k ≥ 0, then the above infinite series converges on the open half-plane of s such that Re(s) > 0.

In both cases f is an analytic function on the corresponding open half plane.

In general the abscissa of convergence of a Dirichlet series is the intercept on the real axis of the vertical line in the complex plane such that there is convergence to the right of it, and divergence to the left. This is the analogue for Dirichlet series of the radius of convergence for power series. The Dirichlet series case is more complicated, though: absolute convergence and uniform convergence may occur in distinct half-planes.

In many cases, the analytic function associated with a Dirichlet series has an analytic extension to a larger domain.

Abscissa of Convergence

Assume that $\sum_{n=1}^{\infty} a_n n^{-s_0}$ converges for some $s_0 \in \mathbb{C}, Re(s_0) > 0$.

- Then $A(N) = \sum_{n=1}^{N} a_n = o(N^{s_0})$. *Proof*: note that $(n+1)^s - n^s = \int_n^{n+1} sx^{s-1}dx = O(n^{s-1})$. Let $B(N) = \sum_{n=1}^{N} \frac{a_n}{n^{s_0}} = L + o(1)$ where $L = \sum_{n=1}^{\infty} a_n n^{-s_0}$, by summation by parts we have $A(N) = \sum_{n=1}^{N} \frac{a_n}{n^{s_0}} n^{s_0} = B(N)N^{s_0} + \sum_{n=1}^{N-1} B(n)(n^{s_0} - (n+1)^{s_0}) = o(N^{s_0}) + \sum_{n=1}^{N-1} O(n^{s_0 - 1}) = o(n^{s_0})$.
- If $\sum_{n=1}^{\infty} a_n$ diverges, then the number $\sigma = \limsup_{N \to \infty} \frac{\ln |A(N)|}{\ln N} = \inf\{\sigma, A(N) = O(N^{\sigma})\}$ is called the *abscissa of convergence* of the Dirichlet series :

  $$\sum_{n=1}^{\infty} a_n n^{-s} \text{ converges for } Re(s) > \sigma \text{ and diverges for } Re(s) < \sigma$$

  From the definition $\forall \epsilon > 0, A(N) = O(N^{\sigma+\epsilon})$, so that $\sum_{n=1}^{N} a_n n^{-s} = A(N)N^{-s} + \sum_{n=1}^{N-1} A(n)(n^{-s} - (n+1)^{-s}) = O(N^{\sigma+\epsilon-s}) + \sum_{n=1}^{N-1} O(n^{\sigma+\epsilon-s-1})$ which converges as $N \to \infty$ whenever $Re(s) > \sigma$. Hence, for every s such that $\sum_{n=1}^{\infty} a_n n^{-s}$ diverges, we have $\sigma \geq Re(s)$, and this finishes the proof.
- If $\sum_{n=1}^{\infty} a_n$ converges then $f(\sigma + it) = o\left(\frac{1}{\sigma}\right)$ as $\sigma \to 0^+$ and where it is meromorphic $f(s)$ has no poles on $Re(s) = 0$

Formal Dirichlet Series

A formal Dirichlet series over a ring R is associated to a function a from the positive integers to R

$$D(a,s) = \sum_{n=1}^{\infty} a(n)n^{-s}$$

with addition and multiplication defined by

$$D(a,s) + D(b,s) = \sum_{n=1}^{\infty} (a+b)(n)n^{-s}$$

$$D(a,s) \cdot D(b,s) = \sum_{n=1}^{\infty} (a*b)(n)n^{-s}$$

where

$$(a+b)(n) = a(n) + b(n)$$

is the pointwise sum and

$$(a*b)(n) = \sum_{k|n} a(k)b(n/k)$$

is the Dirichlet convolution of a and b.

The formal Dirichlet series form a ring Ω, indeed an R-algebra, with the zero function as additive zero element and the function δ defined by δ(1)=1, δ(n)=0 for n>1 as multiplicative identity. An element of this ring is invertible if a(1) is invertible in R. If R is commutative, so is Ω; if R is an integral domain, so is Ω. The non-zero multiplicative functions form a subgroup of the group of units of Ω.

The ring of formal Dirichlet series over **C** is isomorphic to a ring of formal power series in countably many variables.

Derivatives

Given

$$F(s)=\sum_{n=1}^{\infty}\frac{f(n)}{n^s}$$

it is possible to show that

$$F'(s)=-\sum_{n=1}^{\infty}\frac{f(n)\log(n)}{n^s}$$

assuming the right hand side converges. For a completely multiplicative function f(n), and assuming the series converges for Re(s) > $_0$, then one has that

$$\frac{F'(s)}{F(s)}=-\sum_{n=1}^{\infty}\frac{f(n)\Lambda(n)}{n^s}$$

converges for Re(s) > $_0$. Here, (n) is the von Mangoldt function.

Products

Suppose

$$F(s)=\sum_{n=1}^{\infty}f(n)n^{-s}$$

and

$$G(s)=\sum_{n=1}^{\infty}g(n)n^{-s}.$$

If both F(s) and G(s) are absolutely convergent for s > a and s > b then we have

$$\frac{1}{2T}\int_{-T}^{T}F(a+it)G(b-it)\,dt=\sum_{n=1}^{\infty}f(n)g(n)n^{-a-b}\text{ as }T\sim\infty.$$

If a = b and f(n) = g(n) we have

$$\frac{1}{2T}\int_{-T}^{T}|F(a+it)|^2dt=\sum_{n=1}^{\infty}[f(n)]^2n^{-2a}\text{ as }T\sim\infty.$$

Integral Transforms

The Mellin transform of a Dirichlet series is given by Perron's formula.

Relation to Power Series

The sequence a_n generated by a Dirichlet series generating function corresponding to:

$$\zeta(s)^m = \sum_{n=1}^{\infty} \frac{a_n}{n^s}$$

where (s) is the Riemann zeta function, has the ordinary generating function:

$$\sum_{n=1}^{\infty} a_n x^n = x + \binom{m}{1} \sum_{a=2}^{\infty} x^a + \binom{m}{2} \sum_{a=2}^{\infty} \sum_{b=2}^{\infty} x^{ab} + \binom{m}{3} \sum_{a=2}^{\infty} \sum_{b=2}^{\infty} \sum_{c=2}^{\infty} x^{abc} + \binom{m}{4} \sum_{a=2}^{\infty} \sum_{b=2}^{\infty} \sum_{c=2}^{\infty} \sum_{d=2}^{\infty} x^{abcd} + \ldots$$

Riemann Zeta Function

The Riemann zeta function or Euler–Riemann zeta function, ζ(s), is a function of a complex variable s that analytically continues the sum of the infinite series

$$\zeta(s) = \sum_{n=1}^{\infty} \frac{1}{n^s}$$

for when the real part of s is greater than 1. More general representations of ζ(s) for all s are given below. The Riemann zeta function plays a pivotal role in analytic number theory and has applications in physics, probability theory, and applied statistics.

As a function of a real variable, Leonhard Euler first introduced and studied it in the first half of the eighteenth century without using complex analysis, which was not available at the time. Bernhard Riemann's 1859 article "On the Number of Primes Less Than a Given Magnitude" extended the Euler definition to a complex variable, proved its meromorphic continuation and functional equation, and established a relation between its zeros and the distribution of prime numbers.

The values of the Riemann zeta function at even positive integers were computed by Euler. The first of them, (2), provides a solution to the Basel problem. In 1979 Apéry proved the irrationality of (3). The values at negative integer points, also found by Euler, are rational numbers and play an important role in the theory of modular forms. Many generalizations of the Riemann zeta function, such as Dirichlet series, Dirichlet L-functions and L-functions, are known.

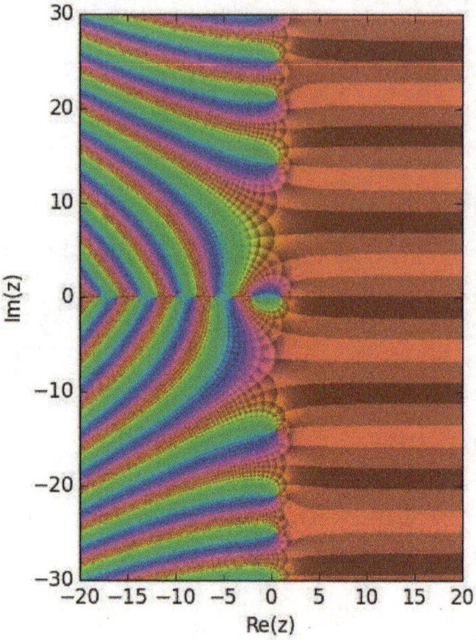

The Riemann zeta function $\zeta(z)$ represented in a rectangular region of the complex plane. It is generated as a Matplotlib plot using a version of the Domain coloring method.

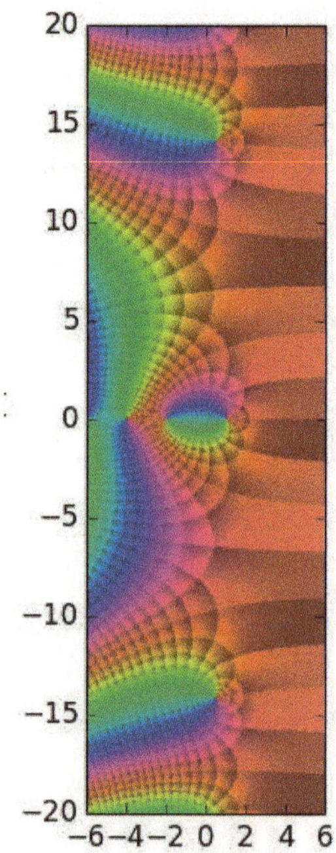

A view of the Riemann zeta function showing the pole z=1, and two zeros on the critical line.

Definition

The Riemann zeta function (s) is a function of a complex variable s = + it. (The notation s, , and t is used traditionally in the study of the -function, following Riemann.)

The following infinite series converges for all complex numbers s with real part greater than 1, and defines (s) in this case:

$$\zeta(s) = \sum_{n=1}^{\infty} n^{-s} = \frac{1}{1^s} + \frac{1}{2^s} + \frac{1}{3^s} + \cdots \qquad \sigma = \Re(s) > 1.$$

It can also be defined by the integral

$$\zeta(s) = \frac{1}{\Gamma(s)} \int_0^{\infty} \frac{x^{s-1}}{e^x - 1} \mathrm{d}x$$

where Γ(s) is the gamma function.

The Riemann zeta function is defined as the analytic continuation of the function defined for σ > 1 by the sum of the preceding series.

Leonhard Euler considered the above series in 1740 for positive integer values of s, and later Chebyshev extended the definition to Re s > 1.

The above series is a prototypical Dirichlet series that converges absolutely to an analytic function for s such that > 1 and diverges for all other values of s. Riemann showed that the function defined by the series on the half-plane of convergence can be continued analytically to all complex values s ≠ 1. For s = 1 the series is the harmonic series which diverges to +∞, and

$$\lim_{s \to 1} (s-1)\zeta(s) = 1.$$

Thus the Riemann zeta function is a meromorphic function on the whole complex s-plane, which is holomorphic everywhere except for a simple pole at s = 1 with residue 1.

Specific Values

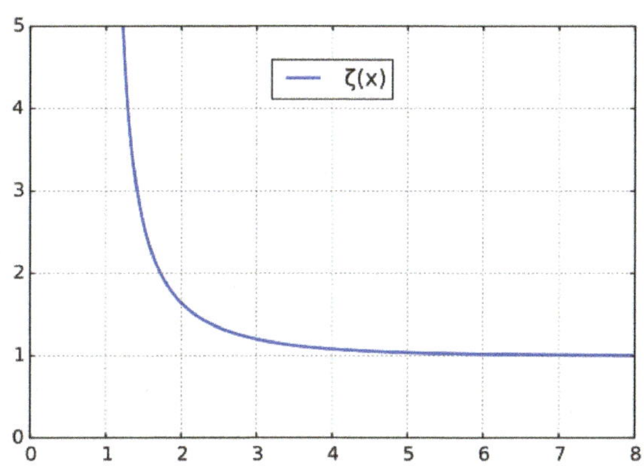

Riemann zeta function for real s > 1

For any positive even integer 2n:

$$\zeta(2n) = \frac{(-1)^{n+1} B_{2n} (2\pi)^{2n}}{2(2n)!}$$

where B_{2n} is a Bernoulli number.

For negative integers, one has

$$\zeta(-n) = -\frac{B_{n+1}}{n+1}$$

for $n \geq 1$ (using the NIST convention that $B_1 = -1/2$), so in particular ζ vanishes at the negative even integers because $B_m = 0$ for all odd m other than 1. For odd positive integers, no such simple expression is known, although these values are thought to be related to the algebraic K-theory of the integers.

Via analytic continuation, one can show that

$$\zeta(-1) = -\frac{1}{12}$$

gives a way to assign a finite result to the divergent series 1 + 2 + 3 + 4 + · · ·, which can be useful in certain contexts such as string theory.

$$\zeta(0) = -\frac{1}{2};$$

$$\zeta(1/2) \approx -1.4603545$$

This is employed in calculating of kinetic boundary layer problems of linear kinetic equations.

$$\zeta(1) = 1 + \frac{1}{2} + \frac{1}{3} + \cdots = \infty;$$

if we approach from numbers larger than 1. Then this is the harmonic series. But its Cauchy principal value

$$\lim_{\varepsilon \to 0} \frac{\zeta(1+\varepsilon) + \zeta(1-\varepsilon)}{2}$$

exists which is the Euler–Mascheroni constant $\gamma = 0.5772\ldots$.

$$\zeta(3/2) \approx 2.612;$$

This is employed in calculating the critical temperature for a Bose–Einstein condensate in a box with periodic boundary conditions, and for spin wave physics in magnetic systems.

$$\zeta(2) = 1 + \frac{1}{2^2} + \frac{1}{3^2} + \cdots = \frac{\pi^2}{6} \approx 1.645;$$

The demonstration of this equality is known as the Basel problem. The reciprocal of this sum answers the question: What is the probability that two numbers selected at random are relatively prime?

$$\zeta(3) = 1 + \frac{1}{2^3} + \frac{1}{3^3} + \cdots \approx 1.202;$$

This is called Apéry's constant.

$$\zeta(4) = 1 + \frac{1}{2^4} + \frac{1}{3^4} + \cdots = \frac{\pi^4}{90} \approx 1.0823;$$

This appears when integrating Planck's law to derive the Stefan–Boltzmann law in physics.

Euler Product Formula

The connection between the zeta function and prime numbers was discovered by Euler, who proved the identity

$$\sum_{n=1}^{\infty} \frac{1}{n^s} = \prod_{p \text{ prime}} \frac{1}{1-p^{-s}},$$

where, by definition, the left hand side is $\zeta(s)$ and the infinite product on the right hand side extends over all prime numbers p (such expressions are called Euler products):

$$\prod_{p \text{ prime}} \frac{1}{1-p^{-s}} = \frac{1}{1-2^{-s}} \cdot \frac{1}{1-3^{-s}} \cdot \frac{1}{1-5^{-s}} \cdot \frac{1}{1-7^{-s}} \cdot \frac{1}{1-11^{-s}} \cdots \frac{1}{1-p^{-s}} \cdots.$$

Both sides of the Euler product formula converge for $\mathrm{Re}(s) > 1$. The proof of Euler's identity uses only the formula for the geometric series and the fundamental theorem of arithmetic. Since the harmonic series, obtained when $s = 1$, diverges, Euler's formula (which becomes $\prod_p p/(p-1)$) implies that there are infinitely many primes.

The Euler product formula can be used to calculate the asymptotic probability that s randomly selected integers are set-wise coprime. Intuitively, the probability that any single number is divisible by a prime (or any integer), p is $1/p$. Hence the probability that s numbers are all divisible by this prime is $1/p^s$, and the probability that at least one of them is not is $1 - 1/p^s$. Now, for distinct primes, these divisibility events are mutually independent because the candidate divisors are coprime (a number is divisible by coprime divisors n and m if and only if it is divisible by nm, an event which occurs with probability $1/(nm)$). Thus the asymptotic probability that s numbers are coprime is given by a product over all primes,

$$\prod_{p\text{ prime}}\left(1-\frac{1}{p^s}\right)=\left(\prod_{p\text{ prime}}\frac{1}{1-p^{-s}}\right)^{-1}=\frac{1}{\zeta(s)}.$$

(More work is required to derive this result formally.)

The Functional Equation

The Riemann zeta function satisfies the functional equation (known as the Riemann functional equation or Riemann's functional equation)

$$\zeta(s)=2^s\pi^{s-1}\ \sin\left(\frac{\pi s}{2}\right)\ \Gamma(1-s)\ \zeta(1-s),$$

where Γ(s) is the gamma function, which is an equality of meromorphic functions valid on the whole complex plane. This equation relates values of the Riemann zeta function at the points s and 1 − s. Owing to the zeros of the sine function, the functional equation implies that (s) has a simple zero at each even negative integer s = −2n — these are known as the trivial zeros of (s). When s is an even positive integer, the product sin(πs/2)Γ(1−s) on the right is regular and non-zero because Γ(1−s) has a simple pole: the functional equation thus relates the values of the Riemann zeta function at odd negative integers and even positive integers.

The functional equation was established by Riemann in his 1859 paper On the Number of Primes Less Than a Given Magnitude and used to construct the analytic continuation in the first place. An equivalent relationship had been conjectured by Euler over a hundred years earlier, in 1749, for the Dirichlet eta function (alternating zeta function)

$$\eta(s)=\sum_{n=1}^{\infty}\frac{(-1)^{n+1}}{n^s}=(1-2^{1-s})\zeta(s).$$

Incidentally, this relation is interesting also because it actually exhibits (s) as a Dirichlet series (of the -function) which is convergent (albeit non-absolutely) in the larger half-plane > 0 (not just > 1), up to an elementary factor.

Riemann also found a symmetric version of the functional equation (which he assigned the letter [small xi]), given by first defining

$$\xi(s)=\frac{1}{2}\pi^{-s/2}s(s-1)\Gamma\left(\frac{s}{2}\right)\zeta(s).$$

The functional equation is then given by

$$\xi(s)=\xi(1-s).$$

(Riemann defined a similar but different function which he called ξ(t).)

Zeros, The Critical Line, and the Riemann Hypothesis

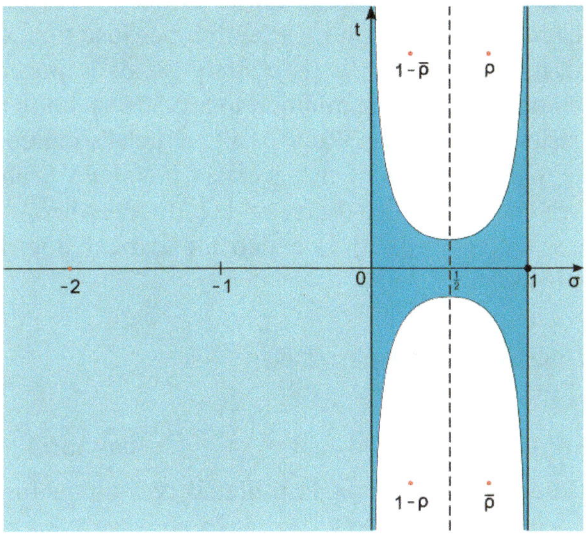

Apart from the trivial zeros, the Riemann zeta function doesn't have any zero on the right of σ=1 and on the left of σ=0 (neither can the zeros lie too close to those lines). Furthermore, the non-trivial zeros are symmetric about the real axis and the line σ = 1/2 and, according to the Riemann Hypothesis, they all lie on the line σ = 1/2.

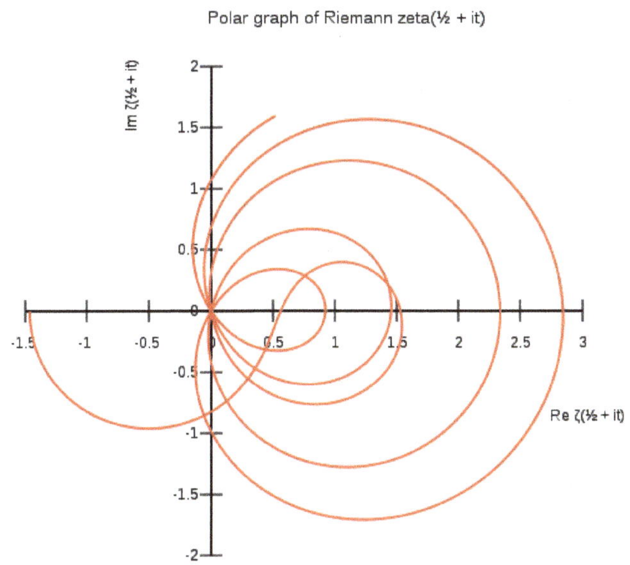

This image shows a plot of the Riemann zeta function along the critical line for real values of t running from 0 to 34. The first five zeros in the critical strip are clearly visible as the place where the spirals pass through the origin.

The functional equation shows that the Riemann zeta function has zeros at −2, −4, ... These are called the trivial **zeros**. They are trivial in the sense that their existence is relatively easy to prove, for example, from sin(πs/2) being 0 in the functional equation. The non-trivial zeros have captured far more attention because their distribution not only is far less understood but, more importantly, their study yields impressive results concerning prime numbers and related objects in number theory. It is known that any non-trivial zero lies in the open strip $\{s \in \mathbf{C} : 0 < \mathrm{Re}(s) < 1\}$, which is called the critical strip. The Riemann hypothesis, considered one of the greatest unsolved problems in mathematics, asserts that any non-trivial zero s has Re(s) = 1/2. In the theory of the Riemann zeta function, the set $\{s \in C : \mathrm{Re}(s) = 1/2\}$ is called the critical line. For the Riemann zeta function on the critical lin.

The Hardy–Littlewood Conjectures

In 1914, Godfrey Harold Hardy proved that $\zeta(\frac{1}{2}+it)$ has infinitely many real zeros.

Hardy and John Edensor Littlewood formulated two conjectures on the density and distance between the zeros of $\zeta(\frac{1}{2}+it)$ on intervals of large positive real numbers. In the following, $N(T)$ is the total number of real zeros and $N_0(T)$ the total number of zeros of odd order of the function $\zeta(\frac{1}{2}+it)$ lying in the interval $(0,T]$.

1. For any $\varepsilon > 0$, there exists a $T_0(\varepsilon) > 0$ such that when $T \geq T_0(\varepsilon)$ and $H = T^{0.25+\varepsilon}$, the interval $(T, T+H]$ contains a zero of odd order.
2. For any $\varepsilon > 0$, there exists a $T_0(\varepsilon) > 0$ and $c_\varepsilon > 0$ such that the inequality $N_0(T+H) - N_0(T) \geq c_\varepsilon H$ holds when $T \geq T_0(\varepsilon)$ and $H = T^{0.5+\varepsilon}$.

These two conjectures opened up new directions in the investigation of the Riemann zeta function.

Zero-Free Region

The location of the Riemann zeta function's zeros is of great importance in the theory of numbers. The prime number theorem is equivalent to the fact that there are no zeros of the zeta function on the Re(s) = 1 line. A better result that follows from an effective form of Vinogradov's mean-value theorem is that $\zeta(\sigma + it) \neq 0$ whenever $|t| \geq 3$ and

$$\sigma \geq 1 - \frac{1}{57.54(\log|t|)^{2/3}(\log\log|t|)^{1/3}}.$$

The strongest result of this kind one can hope for is the truth of the Riemann hypothesis, which would have many profound consequences in the theory of numbers.

Other Results

It is known that there are infinitely many zeros on the critical line. Littlewood showed that if the sequence (γ_n) contains the imaginary parts of all zeros in the upper half-plane in ascending order, then

$$\lim_{n\to\infty}(\gamma_{n+1}-\gamma_n)=0.$$

The critical line theorem asserts that a positive percentage of the nontrivial zeros lies on the critical line.

In the critical strip, the zero with smallest non-negative imaginary part is 1/2 + i14.13472514.... The fact that $\zeta(s)=\overline{\zeta(\overline{s})}$ for all complex s ≠ 1 implies that the zeros of the Riemann zeta function are symmetric about the real axis. Combining this symmetry with the functional equation, furthermore, one sees that the non-trivial zeros are symmetric about the critical line Re(s) = 1/2.

Various Properties

For sums involving the zeta-function at integer and half-integer values, see rational zeta series.

Reciprocal

The reciprocal of the zeta function may be expressed as a Dirichlet series over the Möbius function μ(n):

$$\frac{1}{\zeta(s)}=\sum_{n=1}^{\infty}\frac{\mu(n)}{n^s}$$

for every complex number s with real part > 1. There are a number of similar relations involving various well-known multiplicative functions; these are given in the article on the Dirichlet series.

The Riemann hypothesis is equivalent to the claim that this expression is valid when the real part of s is greater than 1/2.

Universality

The critical strip of the Riemann zeta function has the remarkable property of universality. This zeta-function universality states that there exists some location on the critical strip that approximates any holomorphic function arbitrarily well. Since holomorphic functions are very general, this property is quite remarkable. The very first proof of universality was provided by Sergei Mikhailovitch Voronin in 1975, which results non-applicable to solve Riemann's Hypothesis. A latter and more accurate proof was given in 2003 by R. Garunkštis. In 2015, A. Quartarone gave an even more accurate and general proof of this theorem. Nevertheless, none of the proofs above are definitive or complete. A definitive proof would imply that Riemann's Hypothesis is true.

Estimates of the Maximum of the Modulus of the Zeta Function

Let the functions $F(T;H)$ and $G(s_0;\Delta)$ be defined by the equalities

$$F(T;H) = \max_{|t-T|\le H} |\zeta(\tfrac{1}{2}+it)|, \quad G(s_0;\Delta) = \max_{|s-s_0|\le \Delta} |\zeta(s)|.$$

Here T is a sufficiently large positive number, $0 < H \ll \ln\ln T$, $s_0 = \sigma_0 + iT$, $\frac{1}{2} \le \sigma_0 \le 1$, $0 < \Delta < \frac{1}{3}$. Estimating the values F and G from below shows, how large (in modulus) values $\zeta(s)$ can take on short intervals of the critical line or in small neighborhoods of points lying in the critical strip $0 \le Re\ s \le 1$.

The case $H \gg \ln\ln T$ was studied by Ramachandra; the case $\Delta > c$, where c is a sufficiently large constant, is trivial.

Karatsuba proved, in particular, that if the values H and Δ exceed certain sufficiently small constants, then the estimates

$$F(T;H) \ge T^{-c_1}, \quad G(s_0;\Delta) \ge T^{-c_2},$$

hold, where c_1, c_2 are certain absolute constants.

The Argument of the Riemann Zeta-Function

The function $S(t) = \frac{1}{\pi} \arg \zeta(\frac{1}{2}+it)$ is called the argument of the Riemann zeta function. Here $\zeta(\frac{1}{2}+it)$ is the increment of an arbitrary continuous branch of $\zeta(s)$ along the broken line joining the points $2, 2+it$ and $\frac{1}{2}+it$ There are some theorems on properties of the function S(t). Among those results are the mean value theorems for S(t) and its first integral $S_1(t) = \int_0^t S(u)du$ on intervals of the real line, and also the theorem claiming that every interval $(T, T+H]$ for $H \ge T^{27/82+\varepsilon}$ contains at least

points where the function S(t) changes sign. Earlier similar results were obtained by Atle Selberg for the case $H \ge T^{1/2+\varepsilon}$.

Representations

Dirichlet Series

An extension of the area of convergence can be obtained by rearranging the original series (Konrad Knopp, Theory of Functions, 1945, p. 51-55). The series

$$\zeta(s)=\frac{1}{s-1}\sum_{n=1}^{\infty}\left(\frac{n}{(n+1)^s}-\frac{n-s}{n^s}\right)$$

converges for $\Re s>0$, while

$$\zeta(s)=\frac{1}{s-1}\sum_{n=1}^{\infty}\frac{n(n+1)}{2}\left(\frac{2n+3+s}{(n+1)^{s+2}}-\frac{2n-1-s}{n^{s+2}}\right)$$

converges even for $\Re s>-1$. In this way, the area of convergence can be extended to $\Re s>-k$ for any $k\in\{1,2,3,\ldots\}$.

Mellin Transform

The Mellin transform of a function $f(x)$ is defined as

$$\int_0^\infty f(x)x^{s-1}\,dx,$$

in the region where the integral is defined. There are various expressions for the zeta-function as a Mellin transform. If the real part of s is greater than one, we have

$$\Gamma(s)\zeta(s)=\int_0^\infty \frac{x^{s-1}}{e^x-1}\,dx,$$

where Γ denotes the Gamma function. By modifying the contour, Riemann showed that

$$2\sin(\pi s)\Gamma(s)\zeta(s)=i\oint_H \frac{(-x)^{s-1}}{e^x-1}\,dx$$

for all s. (H denotes the Hankel contour)

We can also find expressions which relate to prime numbers and the prime number theorem. If π(x) is the prime-counting function, then

$$\log\zeta(s)=s\int_0^\infty \frac{\pi(x)}{x(x^s-1)}\,dx,$$

for values with Re(s) > 1.

A similar Mellin transform involves the Riemann prime-counting function J(x), which counts prime powers p^n with a weight of 1/n, so that

$$J(x)=\sum\frac{\pi(x^{1/n})}{n}.$$

Now we have

$$\log \zeta(s) = s \int_0^\infty J(x) x^{-s-1}\, dx.$$

These expressions can be used to prove the prime number theorem by means of the inverse Mellin transform. Riemann's prime-counting function is easier to work with, and $\pi(x)$ can be recovered from it by Möbius inversion.

Theta Functions

The Riemann zeta function can be given formally by a divergent Mellin transform

$$2\pi^{-s/2}\Gamma(s/2)\zeta(s) = \int_0^\infty \theta(it) t^{s/2-1}\, dt,$$

in terms of Jacobi's theta function

$$\theta(\tau) = \sum_{n=-\infty}^{\infty} \exp(\pi i n^2 \tau).$$

However this integral does not converge for any value of s and so needs to be regularized: this gives the following expression for the zeta function:

$$\pi^{-s/2}\Gamma(s/2)\zeta(s)$$

$$= \frac{1}{s-1} - \frac{1}{s} + \frac{1}{2}\int_0^1 \left(\theta(it) - t^{-1/2}\right) t^{s/2-1}\, dt + \frac{1}{2}\int_1^\infty (\theta(it) - 1) t^{s/2-1}\, dt.$$

Laurent Series

The Riemann zeta function is meromorphic with a single pole of order one at s = 1. It can therefore be expanded as a Laurent series about s = 1; the series development then is

$$\zeta(s) = \frac{1}{s-1} + \sum_{n=0}^{\infty} \frac{(-1)^n}{n!} \gamma_n\, (s-1)^n.$$

The constants γ_n here are called the Stieltjes constants and can be defined by the limit

$$\gamma_n = \lim_{m\to\infty} \left[\left(\sum_{k=1}^{m} \frac{(\log k)^n}{k} \right) - \frac{(\log m)^{n+1}}{n+1} \right].$$

The constant term γ_0 is the Euler–Mascheroni constant.

Integral

For all $s \in \mathbb{C} \setminus \{1\}$ the integral relation (cf. Abel–Plana formula)

$$\zeta(s) = \frac{2^{s-1}}{s-1} - 2^s \int_0^\infty \frac{\sin(s \arctan t)}{(1+t^2)^{s/2}(e^{\pi t}+1)} \, dt,$$

holds true, which may be used for a numerical evaluation of the zeta-function.

Rising Factorial

Another series development using the rising factorial valid for the entire complex plane is

$$\zeta(s) = \frac{s}{s-1} - \sum_{n=1}^{\infty} (\zeta(s+n) - 1) \frac{s(s+1)\cdots(s+n-1)}{(n+1)!}.$$

This can be used recursively to extend the Dirichlet series definition to all complex numbers.

The Riemann zeta function also appears in a form similar to the Mellin transform in an integral over the Gauss–Kuzmin–Wirsing operator acting on x^{s-1}; that context gives rise to a series expansion in terms of the falling factorial.

Hadamard Product

On the basis of Weierstrass's factorization theorem, Hadamard gave the infinite product expansion

$$\zeta(s) = \frac{e^{(\log(2\pi)-1-\gamma/2)s}}{2(s-1)\Gamma(1+s/2)} \prod_{\rho} \left(1 - \frac{s}{\rho}\right) e^{s/\rho},$$

where the product is over the non-trivial zeros ρ of ζ and the letter γ again denotes the Euler–Mascheroni constant. A simpler infinite product expansion is

$$\zeta(s) = \pi^{s/2} \frac{\prod_{\rho} \left(1 - \frac{s}{\rho}\right)}{2(s-1)\Gamma(1+s/2)}.$$

This form clearly displays the simple pole at $s = 1$, the trivial zeros at $-2, -4, \ldots$ due to the gamma function term in the denominator, and the non-trivial zeros at $s = \rho$. (To ensure convergence in the latter formula, the product should be taken over "matching pairs" of zeroes, i.e. the factors for a pair of zeroes of the form ρ and $1 - \rho$ should be combined.)

Logarithmic Derivative on the Critical Strip

$$\pi\frac{dN}{dx}(x) = \frac{1}{2i}\frac{d}{dx}\big(\log(\zeta(1/2+ix)) - \log(\zeta(1/2-ix))\big) - \frac{2}{1+4x^2} - \sum_{n=0}^{\infty}\frac{2n+1/2}{(2n+1/2)^2+x^2}$$

where $\frac{dN(x)}{dx} = \sum_{\rho}\delta(x-\rho)$ is the density of zeros of ζ on the critical strip 0 < Re(s) < 1 (δ is the Dirac delta distribution, and the sum is over the nontrivial zeros ρ of ζ).

Globally Convergent Series

A globally convergent series for the zeta function, valid for all complex numbers s except $s = 1 + \frac{2\pi i n}{\log(2)}$ for some integer n, was conjectured by Konrad Knopp and proven by Helmut Hasse in 1930 (cf. Euler summation):

$$\zeta(s) = \frac{1}{1-2^{1-s}}\sum_{n=0}^{\infty}\frac{1}{2^{n+1}}\sum_{k=0}^{n}\binom{n}{k}\frac{(-1)^k}{(k+1)^s}.$$

The series only appeared in an appendix to Hasse's paper, and did not become generally known until it was rediscovered more than 60 years later.

Hasse also proved the globally converging series

$$\zeta(s) = \frac{1}{s-1}\sum_{n=0}^{\infty}\frac{1}{n+1}\sum_{k=0}^{n}\binom{n}{k}\frac{(-1)^k}{(k+1)^{s-1}}$$

in the same publication. However, recent research indicates that the latter series was discovered earlier, by Joseph Ser in 1926.

Peter Borwein has shown a very rapidly convergent series suitable for high precision numerical calculations. The algorithm, making use of Chebyshev polynomials, is described in the article on the Dirichlet eta function.

Series Representation at Positive Integers via the Primorial

$$\zeta(k) = \frac{2^k}{2^k-1} + \sum_{r=2}^{\infty}\frac{(p_{r-1}\#)^k}{J_k(p_r\#)} \qquad (k = 2, 3, \ldots).$$

Here $p_n\#$ is the primorial sequence and J_k is Jordan's totient function.

Series Representation by the Incomplete Poly-Bernoulli Numbers

The function ζ can be represented, for $\Re(s) > 1$, by the infinite series

$$\zeta(s) = \sum_{n=0}^{\infty} B_{n,\geq 2}^{(s)} \frac{(W_k(-1))^n}{n!},$$

where $k \in \{-1, 0\}$, W_k is the k-th branch of the Lambert W function, and $B_{n,\geq 2}^{(\mu)}$ is an incomplete poly-Bernoulli number.

Applications

The zeta function occurs in applied statistics.

Zeta function regularization is used as one possible means of regularization of divergent series and divergent integrals in quantum field theory. In one notable example, the Riemann zeta-function shows up explicitly in the calculation of the Casimir effect. The zeta function is also useful for the analysis of dynamical systems.

Infinite Series

The zeta function evaluated at positive integers appears in infinite series representations of a number of constants. There are more formulas in the article Harmonic number.

$$1 = \sum_{n=2}^{\infty} (\zeta(n) - 1).$$

In fact the even and odd terms give the two sums $\sum_{n=1}^{\infty} (\zeta(2n) - 1) = \frac{3}{4}$ and $\sum_{n=1}^{\infty} (\zeta(2n+1) - 1) = \frac{1}{4}.$

$$\log 2 = \sum_{n=1}^{\infty} \frac{\zeta(2n) - 1}{n}.$$

$$1 - \gamma = \sum_{n=2}^{\infty} \frac{\zeta(n) - 1}{n}$$

where γ is Euler's constant.

$$\log \pi = \sum_{n=2}^{\infty} \frac{(2(\frac{3}{2})^n - 3)(\zeta(n) - 1)}{n}.$$

$$\frac{\pi}{4} = \sum_{n=2}^{\infty} \frac{\zeta(n) - 1}{n} \Im((1+i)^n - (1+i^n))$$

where J represents the imaginary part of a complex number.

Some zeta series evaluate to more complicated expressions

$$\sum_{n=1}^{\infty} \frac{\zeta(2n)-1}{2^{2n}} = \frac{1}{6}.$$

$$\sum_{n=1}^{\infty} \frac{\zeta(2n)-1}{4^{2n}} = \frac{13}{30} - \frac{\pi}{8}.$$

$$\sum_{n=1}^{\infty} \frac{\zeta(2n)-1}{8^{2n}} = \frac{61}{126} - \frac{\pi}{16}(\sqrt{2}+1).$$

$$\sum_{n=1}^{\infty} \frac{\zeta(2n)-1}{a^{2n}} = \frac{1}{2} + \frac{1}{1-a^2} - \frac{\pi \cot(\pi/a)}{2a} \quad \text{for } |a| > 1.$$

$$\sum_{n=1}^{\infty} (\zeta(4n)-1) = \frac{7}{8} - \frac{\pi}{4}\left(\frac{e^{2\pi}+1}{e^{2\pi}-1}\right).$$

Generalizations

There are a number of related zeta functions that can be considered to be generalizations of the Riemann zeta function. These include the Hurwitz zeta function

$$\zeta(s,q) = \sum_{k=0}^{\infty} \frac{1}{(k+q)^s}$$

(the convergent series representation was given by Helmut Hasse in 1930, cf. Hurwitz zeta function), which coincides with the Riemann zeta function when q = 1 (note that the lower limit of summation in the Hurwitz zeta function is 0, not 1), the Dirichlet L-functions and the Dedekind zeta-function. For other related functions see the articles Zeta function and L-function.

The polylogarithm is given by

$$\mathrm{Li}_s(z) = \sum_{k=1}^{\infty} \frac{z^k}{k^s}$$

which coincides with the Riemann zeta function when z = 1.

The Lerch transcendent is given by

$$\Phi(z,s,q) = \sum_{k=0}^{\infty} \frac{z^k}{(k+q)^s}$$

which coincides with the Riemann zeta function when z = 1 and q = 1 (note that the lower limit of summation in the Lerch transcendent is 0, not 1).

The Clausen function $\mathrm{Cl}_s(\theta)$ that can be chosen as the real or imaginary part of $\mathrm{Li}_s(e^{i\theta})$.

The multiple zeta functions are defined by

$$\zeta(s_1, s_2, \ldots, s_n) = \sum_{k_1 > k_2 > \cdots > k_n > 0} k_1^{-s_1} k_2^{-s_2} \cdots k_n^{-s_n}.$$

One can analytically continue these functions to the n-dimensional complex space. The special values of these functions are called multiple zeta values by number theorists and have been connected to many different branches in mathematics and physics.

REFERENCES

1. Apostol, Tom M. (1976), Introduction to analytic number theory, Undergraduate Texts in Mathematics, New York-Heidelberg: Springer-Verlag, ISBN 978-0-387-90163-3, MR 0434929, Zbl 0335.10001
2. Davenport, Harold (2000), Multiplicative number theory, Graduate Texts in Mathematics, 74 (3rd revised ed.), New York: Springer-Verlag, ISBN 978-0-387-95097-6, MR 1790423
3. Tenenbaum, Gérald (1995), Introduction to Analytic and Probabilistic Number Theory, Cambridge studies in advanced mathematics, 46, Cambridge University Press, ISBN 0-521-41261-7
4. Henry Mann (1976). Addition Theorems: The Addition Theorems of Group Theory and Number Theory (Corrected reprint of 1965 Wiley ed.). Huntington, New York: Robert E. Krieger Publishing Company. ISBN 0-88275-418-1.
5. Nathanson, Melvyn B. (1996). Additive Number Theory: The Classical Bases. Graduate Texts in Mathematics. 164. Springer-Verlag. ISBN 0-387-94656-X. Zbl 0859.11002.
6. Nathanson, Melvyn B. (1996). Additive Number Theory: Inverse Problems and the Geometry of Sumsets. Graduate Texts in Mathematics. 165. Springer-Verlag. ISBN 0-387-94655-1. Zbl 0859.11003.
7. Tao, Terence; Vu, Van (2006). Additive Combinatorics. Cambridge Studies in Advanced Mathematics. 105. Cambridge University Press.
8. Mordell, L. J. (1969). Diophantine equations. Pure and Applied Mathematics. 30. Academic Press. ISBN 0-12-506250-8. Zbl 0188.34503.
9. Schmidt, Wolfgang M. (1991). Diophantine approximations and Diophantine equations. Lecture Notes in Mathematics. 1467. Berlin: Springer-Verlag. ISBN 3-540-54058-X. Zbl 0754.11020.
10. Shorey, T. N.; Tijdeman, R. (1986). Exponential Diophantine equations. Cambridge Tracts in Mathematics. 87. Cambridge University Press. ISBN 0-521-26826-5. Zbl 0606.10011.
11. Smart, Nigel P. (1998). The algorithmic resolution of Diophantine equations. London Mathematical Society Student Texts. 41. Cambridge University Press. ISBN 0-521-64156-X. Zbl 0907.11001.
12. Stillwell, John (2004). Mathematics and its History (Second ed.). Springer Science + Business Media Inc. ISBN 0-387-95336-1.

13. Cashwell, E.D.; Everett, C.J. (1959). "The ring of number-theoretic functions". Pacific J. Math. 9: 975–985. doi:10.2140/pjm.1959.9.975. ISSN 0030-8730. MR 0108510. Zbl 0092.04602.
14. Apostol, Tom M. (1976), Introduction to analytic number theory, Undergraduate Texts in Mathematics, New York-Heidelberg: Springer-Verlag, ISBN 978-0-387-90163-3, MR 0434929, Zbl 0335.10001
15. Hardy, G.H.; Riesz, Marcel (1915). The general theory of Dirichlet's series. Cambridge Tracts in Mathematics. 18. Cambridge University Press.
16. The general theory of Dirichlet's series by G. H. Hardy. Cornell University Library Historical Math Monographs. {Reprinted by} Cornell University Library Digital Collections
17. Gould, Henry W.; Shonhiwa, Temba (2008). "A catalogue of interesting Dirichlet series". Miss. J. Math. Sci. 20 (1).
18. Mathar, Richard J. (2011). "Survey of Dirichlet series of multiplicative arithmetic functions". arXiv:1106.4038Freely accessible [math.NT].
19. Tenenbaum, Gérald (1995). Introduction to Analytic and Probabilistic Number Theory. Cambridge Studies in Advanced Mathematics. 46. Cambridge University Press. ISBN 0-521-41261-7. Zbl 0831.11001.
20. Apostol, T. M. (2010), "Zeta and Related Functions", in Olver, Frank W. J.; Lozier, Daniel M.; Boisvert, Ronald F.; Clark, Charles W., NIST Handbook of Mathematical Functions, Cambridge University Press, ISBN 978-0521192255, MR2723248
21. Jonathan Borwein, David M. Bradley, Richard Crandall (2000). "Computational Strategies for the Riemann Zeta Function" (PDF). J. Comp. App. Math. 121 (1–2): 247–296. Bibcode:2000JCoAM.121..247B. doi:10.1016/S0377-0427(00)00336-8. Bibcode: 2000JCoAM.121..247B
22. Cvijović, Djurdje; Klinowski, Jacek (2002). "Integral Representations of the Riemann Zeta Function for Odd-Integer Arguments". J. Comp. App. Math. 142 (2): 435–439. Bibcode:2002JCoAM.142..435C. doi:10.1016/S0377-0427(02)00358-8. MR 1906742. Bibcode: 2002JCoAM.142..435C
23. Cvijović, Djurdje; Klinowski, Jacek (1997). "Continued-fraction expansions for the Riemann zeta function and polylogarithms". Proc. Amer. Math. Soc. 125 (9): 2543–2550. doi:10.1090/S0002-9939-97-04102-6.
24. H. M. Edwards (1974). Riemann's Zeta Function. Academic Press. ISBN 0-486-41740-9. Has an English translation of Riemann's paper.
25. Hadamard, Jacques (1896). "Sur la distribution des zéros de la fonction ζ(s) et ses conséquences arithmétiques". Bulletin de la Societé Mathématique de France. 14: 199–220.
26. G. H. Hardy (1949). Divergent Series. Clarendon Press, Oxford.
27. Hasse, Helmut (1930). "Ein Summierungsverfahren für die Riemannsche ζ-Reihe". Math. Z. 32: 458–464. doi:10.1007/BF01194645. MR 1545177. (Globally convergent series expression.)

28. Ivic (1985). The Riemann Zeta Function. John Wiley & Sons. ISBN 0-471-80634-X.

29. Y. Motohashi (1997). Spectral Theory of the Riemann Zeta-Function. Cambridge University Press. ISBN 0521445205.

30. A.A. Karatsuba; S.M. Voronin (1992). The Riemann Zeta-Function. W. de Gruyter, Berlin.

31. Mező, István; Dil, Ayhan (2010). "Hyperharmonic series involving Hurwitz zeta function". Journal of Number Theory. 130 (2): 360–369. doi:10.1016/j.jnt.2009.08.005. MR 2564902.

32. Hugh L. Montgomery; Robert C. Vaughan (2007). Multiplicative number theory I. Classical theory. Cambridge tracts in advanced mathematics. 97. Cambridge University Press. ISBN 0-521-84903-9. Chapter 10.

33. Donald J. Newman (1998). Analytic number theory. GTM. 177. Springer-Verlag. ISBN 0-387-98308-2. Chapter 6.

34. Raoh, Guo (1996). "The Distribution of the Logarithmic Derivative of the Riemann Zeta Function". Proceedings of the London Mathematical Society. s3–72: 1–27. doi:10.1112/plms/s3-72.1.1.

35. Riemann, Bernhard (1859). "Über die Anzahl der Primzahlen unter einer gegebenen Grösse". Monatsberichte der Berliner Akademie.. In Gesammelte Werke, Teubner, Leipzig (1892), Reprinted by Dover, New York (1953).

36. Sondow, Jonathan (1994). "Analytic continuation of Riemann's zeta function and values at negative integers via Euler's transformation of series". Proc. Amer. Math. Soc. 120 (120): 421–424. doi:10.1090/S0002-9939-1994-1172954-7.

37. E. C. Titchmarsh (1986). The Theory of the Riemann Zeta Function, Second revised (Heath-Brown) edition. Oxford University Press.

38. E. T. Whittaker and G. N. Watson (1927). A Course in Modern Analysis, fourth edition, Cambridge University Press (Chapter XIII).

39. Zhao, Jianqiang (1999). "Analytic continuation of multiple zeta functions". Proc. Amer. Math. Soc. 128 (5): 1275–1283. doi:10.1090/S0002-9939-99-05398-8. MR 1670846.

CHAPTER 5

Geometry of Numbers

Chapter Contents

The branch of number theory that studies number-theoretical problems by the use of geometric methods. Geometry of numbers in its proper sense was formulated by H. Minkowski in 1896 in his fundamental monograph. The starting point of this science, which subsequently became an independent branch of number theory, is the fact (already noted by Minkowski) that certain assertions which seem evident in the context of figures in an -dimensional Euclidean space have far-reaching consequences in number theory.

INTRODUCTION

In number theory, the geometry of numbers studies convex bodies and integer vectors in n-dimensional space. The geometry of numbers was initiated by Hermann Minkowski (1910).

The geometry of numbers has a close relationship with other fields of mathematics, especially functional analysis and Diophantine approximation, the problem of finding rational numbers that approximate an irrational quantity.

MINKOWSKI'S THEOREM

In mathematics, Minkowski's theorem is the statement that any convex set in R^n which is symmetric with respect to the origin and with volume greater than 2^n d(L) contains a non-zero lattice point. The theorem was proved by Hermann Minkowski in 1889 and became the foundation of the branch of number theory called the geometry of numbers.

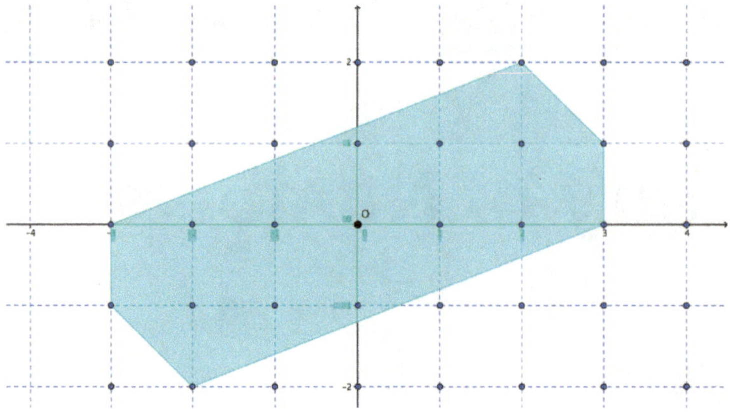

A set in $\mathbf{R}^2$ satisfying the hypotheses of Minkowski's theorem.

Formulation

Suppose that L is a lattice of determinant $d(L)$ in the n-dimensional real vector space $\mathbf{R}^n$ and S is a convex subset of $\mathbf{R}^n$ that is symmetric with respect to the origin, meaning that if x is in S then $-x$ is also in S. Minkowski's theorem states that if the volume of S is strictly greater than $2^n\, d(L)$, then S must contain at least one lattice point other than the origin.

Example

The simplest example of a lattice is the set $\mathbf{Z}^n$ of all points with integer coefficients; its determinant is 1. For $n = 2$, the theorem claims that a convex figure in the plane symmetric about the origin and with area greater than 4 encloses at least one lattice point in addition to the origin. The area bound is sharp: if S is the interior of the square with vertices $(\pm 1, \pm 1)$ then S is symmetric and convex, has area 4, but the only lattice point it contains is the origin. This observation generalizes to every dimension n.

Proof

The following argument proves Minkowski's theorem for the special case of $\mathbf{L}=\mathbf{Z}^2$. It can be generalized to arbitrary lattices in arbitrary dimensions.

Consider the map $f : S \to \mathbb{R}^2, (x, y) \mapsto (x \bmod 2, y \bmod 2)$. Intuitively, this map cuts the plane into 2 by 2 squares, then stacks the squares on top of each other. Clearly f(S) has area ≤ 4. Suppose f were injective, which means the pieces of S cut out by the squares stack up in a non-overlapping way. Since f is locally area-preserving, this non-overlapping property would make it area-preserving for all of S, so the area of $f(S)$ would be the same as that of S, which is greater than 4. That is not the case, so f is not injective, and $f(p_1) = f(p_2)$ for some pair of points p_1, p_2 in S. Moreover, we know from the definition of **f** that $p_2 = p_1 + (2i, 2j)$ for some integers i and j, where i and j are not both zero.

Then since S is symmetric about the origin, $-p_1$ is also a point in S. Since S is convex, the line segment between $-p_1$ and p_2 lies entirely in S, and in particular the midpoint of that segment lies in S. In other words,

$$\frac{1}{2}(-p_1 + p_2) = \frac{1}{2}(-p_1 + p_1 + (2i, 2j)) = (i, j)$$

lies in S. (i,j) is a lattice point, and is not the origin since i and j are not both zero, and so we have found the point we're looking for.

Applications

An application of this theorem is the result that every class in the ideal class group of a number field K contains an integral ideal of norm not exceeding a certain bound, depending on K, called Minkowski's bound: the finiteness of the class number of an algebraic number field follows immediately.

Minkowski's theorem is also useful to prove Lagrange's four-square theorem, which states that every natural number can be written as the sum of the squares of four natural numbers.

MINKOWSKI'S SECOND THEOREM

In mathematics, Minkowski's second theorem is a result in the Geometry of numbers about the values taken by a norm on a lattice and the volume of its fundamental cell.

Setting

Let K be a closed convex centrally symmetric body of positive finite volume in n-dimensional Euclidean space $\mathbf{R}^n$. The *gauge* or *distance* Minkowski functional g attached to K is defined by

$$g(x) = \inf\{\lambda \in \mathbb{R} : x \in \lambda K\}.$$

Conversely, given a norm g on $\mathbf{R}^n$ we define K to be

$$K = \{x \in \mathbb{R}^n : g(x) \leq 1\}.$$

Let Γ be a lattice in $\mathbf{R}^n$. The successive minima of K or g on Γ are defined by setting the k-th successive minimum λ_k to be the infimum of the numbers λ such that λK contains k linearly independent vectors of Γ. We have $0 < \lambda_1 \leq \lambda_2 \leq ... \leq \lambda_n < \infty$.

Statement of the Theorem

The successive minima satisfy

$$\frac{2^n}{n!}\mathrm{vol}(\mathbb{R}^n/\Gamma) \le \lambda_1\lambda_2\cdots\lambda_n\mathrm{vol}(K) \le 2^n\mathrm{vol}(\mathbb{R}^n/\Gamma).$$

Later Research in the Geometry of Numbers

In 1930-1960 research on the geometry of numbers was conducted by many number theorists (including Louis Mordell, Harold Davenport and Carl Ludwig Siegel). In recent years, Lenstra, Brion, and Barvinok have developed combinatorial theories that enumerate the lattice points in some convex bodies.

Subspace Theorem of W. M. Schmidt

In mathematics, the subspace theorem is a result obtained by Wolfgang M. Schmidt (1972). It states that if $L_1,\ldots,L_n$ are linearly independent linear forms in n variables with algebraic coefficients and if $\varepsilon>0$ is any given real number, then the non-zero integer points x with

$$|L_1(x)\cdots L_n(x)| < |x|^{-\epsilon}$$

lie in a finite number of proper subspaces of $\mathbf{Q}^n$.

A quantitative form of the theorem, in which the number of subspaces containing all solutions, was also obtained by Schmidt, and the theorem was generalised by Schlickewei (1977) to allow more general absolute values on number fields.

The theorem may be used to obtain results on Diophantine equations such as Siegel's theorem on integral points and solution of the S-unit equation.

A corollary on Diophantine approximation

The following corollary to the subspace theorem is often itself referred to as the *subspace theorem*. If $a_1,\ldots,a_n$ are algebraic such that $1,a_1,\ldots,a_n$ are linearly independent over $\mathbf{Q}$ and $\varepsilon>0$ is any given real number, then there are only finitely many rational n-tuples $(x_1/y,\ldots,x_n/y)$ with

$$|a_i - x_i/y| < y^{-(1+1/n+\epsilon)}, \quad i = 1,\ldots,n.$$

The specialization $n = 1$ gives the Thue–Siegel–Roth theorem. One may also note that the exponent $1+1/n+\varepsilon$ is best possible by Dirichlet's theorem on diophantine approximation.

INFLUENCE ON FUNCTIONAL ANALYSIS

Minkowski's geometry of numbers had a profound influence on functional analysis. Minkowski proved that symmetric convex bodies induce norms in finite-dimensional vector spaces. Minkowski's theorem was generalized to topological vector spaces by Kolmogorov, whose theorem states that the symmetric convex sets that are closed and bounded generate the topology of a Banach space.

Researchers continue to study generalizations to star-shaped sets and other non-convex sets.

Normed Vector Space

In mathematics, a normed vector space is a vector space on which a norm is defined. In a vector space with 2- or 3-dimensional vectors with real-valued entries, the idea of the "length" of a vector is intuitive. This intuition can easily be extended to any real vector space $\mathbf{R}^n$. The length of a vector in such a vector space has the following properties:

1. The zero vector, 0, has zero length; every other vector has a positive length.

   $\|x\| \geq 0$, and $\|x\| = 0$ if and only if $x = 0$

2. Multiplying a vector by a positive number changes its length without changing its direction. Moreover,

   $\|\alpha x\| = |\alpha| \|x\|$ for any scalar α.

3. The triangle inequality holds. That is, taking norms as distances, the distance from point A through B to C is never shorter than going directly from A to C, or the shortest distance between any two points is a straight line.

   $\|x + y\| \leq \|x\| + \|y\|$ for any vectors x and y. (triangle inequality)

The generalization of these three properties to more abstract vector spaces leads to the notion of **norm**. A vector space on which a norm is defined is then called a normed space or normed vector space. Normed vector spaces are central to the study of linear algebra and functional analysis.

Definition

A normed vector space is a pair (V, $\|\cdot\|$) where V is a vector space and $\|\cdot\|$ a norm on V.

A seminormed vector space is a pair (V,p) where V is a vector space and p a seminorm on V.

We often omit p or $\|\cdot\|$ and just write V for a space if it is clear from the context what (semi) norm we are using.

In a more general sense, a vector norm can be taken to be any real-valued function that satisfies these three properties. The properties 1. and 2. together imply that

$\|x\| = 0$ if and only if $x = 0$.

A useful variation of the triangle inequality is

$\|x - y\| \geq |\|x\| - \|y\||$ for any vectors x and y.

This also shows that a vector norm is a continuous function.

Note that property 2 depends on a choice of norm $|\alpha|$ on the field of scalars. When the scalar field is R (or more generally a subset of C), this is usually taken to be the ordinary absolute value, but other choices are possible. For example, for a vector space over $\mathbb{Q}$ one could take $|\alpha|$ to be the p-adic norm, which gives rise to a different class of normed vector spaces.

Topological Structure

If $(V, \|\cdot\|)$ is a normed vector space, the norm $\|\cdot\|$ induces a metric (a notion of *distance*) and therefore a topology on V. This metric is defined in the natural way: the distance between two vectors **u** and **v** is given by $\|u-v\|$. This topology is precisely the weakest topology which makes $\|\cdot\|$ continuous and which is compatible with the linear structure of V in the following sense:

- The vector addition $+ : V \times V \to V$ is jointly continuous with respect to this topology. This follows directly from the triangle inequality.
- The scalar multiplication $\cdot : \mathbf{K} \times V \to V$, where $\mathbf{K}$ is the underlying scalar field of V, is jointly continuous. This follows from the triangle inequality and homogeneity of the norm.

Similarly, for any semi-normed vector space we can define the distance between two vectors **u** and **v** as $\|u-v\|$. This turns the seminormed space into a pseudometric space (notice this is weaker than a metric) and allows the definition of notions such as continuity and convergence. To put it more abstractly every semi-normed vector space is a topological vector space and thus carries a topological structure which is induced by the semi-norm.

Of special interest are complete normed spaces called Banach spaces. Every normed vector space V sits as a dense subspace inside a Banach space; this Banach space is essentially uniquely defined by V and is called the *completion* of V.

All norms on a finite-dimensional vector space are equivalent from a topological viewpoint as they induce the same topology (although the resulting metric spaces need not be the same).And since any Euclidean space is complete, we can thus conclude that all finite-dimensional normed vector spaces are Banach spaces. A normed vector space V is locally compact if and only if the unit ball $B = \{x : \|x\| \leq 1\}$ is compact, which is the case if and only if V is finite-dimensional; this is a consequence of Riesz's lemma. (In fact, a more general result is true: a topological vector space is locally compact if and only if it is finite-dimensional. The point here is that we don't assume the topology comes from a norm.)

The topology of a seminormed vector space has many nice properties. Given a neighborhood system

$$\mathcal{N}(x) = x + \mathcal{N}(0) := \{x + N \mid N \in \mathcal{N}(0)\}$$

with

$$x + N := \{x + n \mid n \in N\}.$$

Moreover there exists a neighbourhood basis for 0 consisting of absorbing and convex sets. As this property is very useful in functional analysis, generalizations of normed vector spaces with this property are studied under the name locally convex spaces.

Linear Maps and Dual Spaces

The most important maps between two normed vector spaces are the continuous linear maps. Together with these maps, normed vector spaces form a category.

The norm is a continuous function on its vector space. All linear maps between finite dimensional vector spaces are also continuous.

An *isometry* between two normed vector spaces is a linear map f which preserves the norm (meaning $\|f(v)\| = \|v\|$ for all vectors **v**). Isometries are always continuous and injective. A surjective isometry between the normed vector spaces V and W is called an *isometric isomorphism*, and V and W are called *isometrically isomorphic*. Isometrically isomorphic normed vector spaces are identical for all practical purposes.

When speaking of normed vector spaces, we augment the notion of dual space to take the norm into account. The dual V ' of a normed vector space V is the space of all *continuous* linear maps from V to the base field (the complexes or the reals) — such linear maps are called "functionals". The norm of a functional φ is defined as the supremum of $|\varphi(\mathbf{v})|$ where **v** ranges over all unit vectors (i.e. vectors of norm 1) in V. This turns V ' into a normed vector space. An important theorem about continuous linear functionals on normed vector spaces is the Hahn–Banach theorem.

Normed Spaces as Quotient Spaces of Seminormed Spaces

The definition of many normed spaces (in particular, Banach spaces) involves a seminorm defined on a vector space and then the normed space is defined as the quotient space by the subspace of elements of seminorm zero. For instance, with the L^p spaces, the function defined by

$$\|f\|_p = \left(\int |f(x)|^p \, dx\right)^{1/p}$$

is a seminorm on the vector space of all functions on which the Lebesgue integral on the right hand side is defined and finite. However, the seminorm is equal to zero for any function supported on a set of Lebesgue measure zero. These functions form a subspace which we "quotient out", making them equivalent to the zero function.

Finite Product Spaces

Given n seminormed spaces X_i with seminorms q_i we can define the product space as

$$X := \prod_{i=1}^{n} X_i$$

with vector addition defined as

$$(x_1, \ldots, x_n) + (y_1, \ldots, y_n) := (x_1 + y_1, \ldots, x_n + y_n)$$

and scalar multiplication defined as

$$\alpha(x_1, \ldots, x_n) := (\alpha x_1, \ldots, \alpha x_n).$$

We define a new function q

$$q : X \mapsto \mathbb{R}$$

for example as

$$q : (x_1, \ldots, x_n) \to \sum_{i=1}^{n} q_i(x_i).$$

which is a seminorm on X. The function q is a norm if and only if all q_i are norms. More generally, for each real $p \geq 1$ we have the seminorm:

$$q : (x_1, \ldots, x_n) \to \left(\sum_{i=1}^{n} q_i(x_i)^p \right)^{\frac{1}{p}}.$$

For each p this defines the same topological space.

A straightforward argument involving elementary linear algebra shows that the only finite-dimensional seminormed spaces are those arising as the product space of a normed space and a space with trivial seminorm. Consequently, many of the more interesting examples and applications of seminormed spaces occur for infinite-dimensional vector spaces.

BANACH SPACE

In mathematics, more specifically in functional analysis, a Banach space is a complete normed vector space. Thus, a Banach space is a vector space with a metric that allows the computation of vector length and distance between vectors and is complete in the sense that a Cauchy sequence of vectors always converges to a well defined limit that is within the space.

Banach spaces are named after the Polish mathematician Stefan Banach, who introduced this concept and studied it systematically in 1920–1922 along with Hans Hahn and Eduard Helly. Banach spaces originally grew out of the study of function spaces by Hilbert, Fréchet, and Riesz earlier in the century. Banach spaces play a central role in functional analysis. In other areas of analysis, the spaces under study are often Banach spaces.

Definition

A Banach space is a vector space X over the field **R** of real numbers, or over the field **C** of complex numbers, which is equipped with a norm and which is complete with respect to that norm, that is to say, for every Cauchy sequence $\{x_n\}$ in X, there exists an element x in X such that

$$\lim_{n\to\infty} x_n = x,$$

or equivalently:

$$\lim_{n\to\infty} \|x_n - x\|_X = 0.$$

The vector space structure allows one to relate the behavior of Cauchy sequences to that of converging series of vectors. A normed space X is a Banach space if and only if each absolutely convergent series in X converges,

$$\sum_{n=1}^{\infty} \|v_n\|_X < \infty \quad \text{implies that} \quad \sum_{n=1}^{\infty} v_n \text{ converges in } X.$$

Completeness of a normed space is preserved if the given norm is replaced by an equivalent one.

All norms on a finite-dimensional vector space are equivalent. Every finite-dimensional normed space over **R** or **C** is a Banach space.

General Theory

Linear Operators, Isomorphisms

Bounded operator

If X and Y are normed spaces over the same ground field **K**, the set of all continuous **K**-linear maps $T : X \to Y$ is denoted by $B(X, Y)$. In infinite-dimensional spaces, not all linear maps are continuous. A linear mapping from a normed space X to another normed space is continuous if and only if it is bounded on the closed unit ball of X. Thus, the vector space $B(X, Y)$can be given the operator norm

$$\|T\| = \sup\{\|Tx\|_Y \mid x \in X, \ \|x\|_X \le 1\}.$$

For Y a Banach space, the space $B(X, Y)$ is a Banach space with respect to this norm.

If X is a Banach space, the space $B(X) = B(X, X)$ forms a unital Banach algebra; the multiplication operation is given by the composition of linear maps.

If X and Y are normed spaces, they are isomorphic normed spaces if there exists a linear bijection $T : X \rightarrow Y$ such that T and its inverse T^{-1} are continuous. If one of the two spaces X or Y is complete (or reflexive, separable, etc.) then so is the other space. Two normed spaces X and Y are isometrically isomorphic if in addition, T is an isometry, i.e., $\|T(x)\| = \|x\|$for every x in X. The Banach-Mazur distance $d(X, Y)$ between two isomorphic but not isometric spaces X and Y gives a measure of how much the two spaces X and Y differ.

Basic Notions

Every normed space X can be isometrically embedded in a Banach space. More precisely, there is a Banach space Y and an isometric mapping $T : X \rightarrow Y$ such that $T(X)$ is dense in Y. If Z is another Banach space such that there is an isometric isomorphism from X onto a dense subset of Z, then Z is isometrically isomorphic to Y.

This Banach space Y is the completion of the normed space X. The underlying metric space for Y is the same as the metric completion of X, with the vector space operations extended from X to Y. The completion of X is often denoted by $\widehat{X}$.

The cartesian product $X \times Y$ of two normed spaces is not canonically equipped with a norm. However, several equivalent norms are commonly used, such as

$$\|(x, y)\|_1 = \|x\| + \|y\|, \qquad \|(x, y)\|_\infty = \max(\|x\|, \|y\|)$$

and give rise to isomorphic normed spaces. In this sense, the product $X \times Y$ (or the direct sum $X \oplus Y$) is complete if and only if the two factors are complete.

If M is a closed linear subspace of a normed space X, there is a natural norm on the quotient space X/M,

$$\|x + M\| = \inf_{m \in M} \|x + m\|.$$

The quotient X/M is a Banach space when X is complete. The quotient map from X onto X/M, sending x in X to its class $x + M$, is linear, onto and has norm 1, except when $M = X$, in which case the quotient is the null space.

The closed linear subspace M of X is said to be a complemented subspace of X if M is the range of a bounded linear projection P from X onto M. In this case, the space X is isomorphic to the direct sum of M and Ker(P), the kernel of the projection P.

Suppose that X and Y are Banach spaces and that $T \in B(X, Y)$. There exists a canonical factorization of T as

$$T = T_1 \circ \pi, \quad T : X \xrightarrow{\pi} X/\operatorname{Ker}(T) \xrightarrow{T_1} Y$$

where the first map ϖ is the quotient map, and the second map T_1 sends every class x + Ker(T) in the quotient to the image $T(x)$ in Y. This is well defined because all elements in the same class have the same image. The mapping T_1 is a linear bijection from $X/$Ker(T) onto the range $T(X)$, whose inverse need not be bounded.

Classical Spaces

Basic examples of Banach spaces include: the L^p spaces and their special cases, the sequence spaces ℓ^p that consist of scalar sequences indexed by **N**; among them, the space ℓ^1 of absolutely summable sequences and the space ℓ^2 of square summable sequences; the space c_0 of sequences tending to zero and the space ℓ^∞ of bounded sequences; the space $C(K)$ of continuous scalar functions on a compact Hausdorff space K, equipped with the max norm,

$$\|f\|_{C(K)} = \max\{|f(x)| : x \in K\}, \quad f \in C(K).$$

According to the Banach–Mazur theorem, every Banach space is isometrically isomorphic to a subspace of some $C(K)$. For every separable Banach space X, there is a closed subspace M of ℓ^1 such that $X \cong \ell^1/M$.

Any Hilbert space serves as an example of a Banach space. A Hilbert space H on $\mathbf{K} = \mathbf{R}, \mathbf{C}$ is complete for a norm of the form

$$\|x\|_H = \sqrt{\langle x, x\rangle},$$

where

$$\langle\cdot,\cdot\rangle : H \times H \to \mathbf{K}$$

is the inner product, linear in its first argument that satisfies the following:

$$\begin{aligned} \forall x, y \in H : \quad & \langle y, x\rangle = \overline{\langle x, y\rangle}, \\ \forall x \in H : \quad & \langle x, x\rangle \geq 0, \\ & \langle x, x\rangle = 0 \Leftrightarrow x = 0. \end{aligned}$$

For example, the space L^2 is a Hilbert space.

The Hardy spaces, the Sobolev spaces are examples of Banach spaces that are related to L^p spaces and have additional structure. They are important in different branches of analysis, Harmonic analysis and Partial differential equations among others.

Banach Algebras

A Banach algebra is a Banach space A over K = R or C, together with a structure of algebra over K, such that the product map $(a, b) \in A \times A \to ab \in A$ is continuous. An equivalent norm on A can be found so that $\|ab\| \leq \|a\| \|b\|$ for all $a, b \in A$.

Examples

- The Banach space $C(K)$, with the pointwise product, is a Banach algebra.
- The disk algebra $A(\mathbf{D})$ consists of functions holomorphic in the open unit disk $\mathbf{D} \subset \mathbf{C}$ and continuous on its closure: $\mathbf{D}$. Equipped with the max norm on $\mathbf{D}$,

the disk algebra $A(\mathbf{D})$ is a closed subalgebra of $C(\mathbf{D})$.

- The Wiener algebra $A(\mathbf{T})$ is the algebra of functions on the unit circle $\mathbf{T}$ with absolutely convergent Fourier series. Via the map associating a function on $\mathbf{T}$ to the sequence of its Fourier coefficients, this algebra is isomorphic to the Banach algebra $\ell^1(\mathbf{Z})$, where the product is the convolution of sequences.
- For every Banach space X, the space $B(X)$ of bounded linear operators on X, with the composition of maps as product, is a Banach algebra.
- A C*-algebra is a complex Banach algebra A with an antilinear involution $a \to a^*$ such that $\|a^*a\| = \|a\|^2$. The space $B(H)$ of bounded linear operators on a Hilbert space H is a fundamental example of C*-algebra. The Gelfand–Naimark theorem states that every C*-algebra is isometrically isomorphic to a C*-subalgebra of some $B(H)$. The space $C(K)$ of complex continuous functions on a compact Hausdorff space K is an example of commutative C*-algebra, where the involution associates to every function f its complex conjugate $\bar{f}$.

Dual Space

If X is a normed space and $\mathbf{K}$ the underlying field (either the real or the complex numbers), the **continuous dual space** is the space of continuous linear maps from X into $\mathbf{K}$, or **continuous linear functionals**. The notation for the continuous dual is $X' = B(X, \mathbf{K})$ in this article. Since $\mathbf{K}$ is a Banach space (using the absolute value as norm), the dual X' is a Banach space, for every normed space X.

The main tool for proving the existence of continuous linear functionals is the Hahn–Banach theorem.

Hahn–Banach theorem. Let X be a vector space over the field K = R, C. Let further

- $Y \subseteq X$ be a linear subspace,
- $p : X \to \mathbf{R}$ be a sublinear function and
- $f : Y \to \mathbf{K}$ be a linear functional so that $\operatorname{Re}(f(y)) \le p(y)$ for all y in Y.
- Then, there exists a linear functional $F : X \to \mathbf{K}$ so that

$$F|_Y = f, \quad \text{and} \quad \forall x \in X, \ \ \operatorname{Re}(F(x)) \le p(x).$$

In particular, every continuous linear functional on a subspace of a normed space can be continuously extended to the whole space, without increasing the norm of the functional. An important special case is the following: for every vector x in a normed space X, there exists a continuous linear functional f on X such that

$$f(x) = \|x\|_X, \quad \|f\|_{X'} \le 1.$$

When x is not equal to the $\mathbf{0}$ vector, the functional f must have norm one, and is called a norming functional for x.

The Hahn–Banach separation theorem states that two disjoint non-empty convex sets in a real Banach space, one of them open, can be separated by a closed affine hyperplane. The open convex set lies strictly on one side of the hyperplane, the second convex set lies on the other side but may touch the hyperplane.

A subset S in a Banach space X is **total** if the linear span of S is dense in X. The subset S is total in X if and only if the only continuous linear functional that vanishes on S is the **0**functional: this equivalence follows from the Hahn–Banach theorem.

If X is the direct sum of two closed linear subspaces M and N, then the dual X' of X is isomorphic to the direct sum of the duals of M and N. If M is a closed linear subspace in X, one can associate the *orthogonal of M* in the dual,

$$M^{\perp} = \{x' \in X' : x'(m) = 0, \ \forall m \in M\}.$$

The orthogonal $M^{\perp}$ is a closed linear subspace of the dual. The dual of M is isometrically isomorphic to $X'/M^{\perp}$. The dual of X/M is isometrically isomorphic to $M^{\perp}$.

The dual of a separable Banach space need not be separable, but:

Theorem. Let X be a normed space. If X' is separable, then X is separable.

When X' is separable, the above criterion for totality can be used for proving the existence of a countable total subset in X.

Weak Topologies

The weak topology on a Banach space X is the coarsest topology on X for which all elements x' in the continuous dual space X' are continuous. The norm topology is therefore finer than the weak topology. It follows from the Hahn–Banach separation theorem that the weak topology is Hausdorff, and that a norm-closed convex subset of a Banach space is also weakly closed. A norm-continuous linear map between two Banach spaces X and Y is also weakly continuous, i.e., continuous from the weak topology of X to that of Y.

If X is infinite-dimensional, there exist linear maps which are not continuous. The space X^* of all linear maps from X to the underlying field **K** (this space X^* is called the algebraic dual space, to distinguish it from X') also induces a topology on X which is finer than the weak topology, and much less used in functional analysis.

On a dual space X', there is a topology weaker than the weak topology of X', called **weak* topology**. It is the coarsest topology on X' for which all evaluation maps $x' \in X' \to x'(x)$, $x \in X$, are continuous. Its importance comes from the Banach–Alaoglu theorem.

Banach–Alaoglu Theorem. Let X be a normed vector space. Then the closed unit ball $B' = \{x' \in X' : \|x'\| \leq 1\}$ of the dual space is compact in the weak* topology.

The Banach–Alaoglu theorem depends on Tychonoff's theorem about infinite products of compact spaces. When X is separable, the unit ball B' of the dual is a metrizable compact in the weak* topology.

Examples of Dual Spaces

The dual of c_0 is isometrically isomorphic to ℓ^1: for every bounded linear functional f on c_0, there is a unique element $y = \{y_n\} \in \ell^1$ such that

$$f(x) = \sum_{n \in \mathbf{N}} x_n y_n, \qquad x = \{x_n\} \in c_0, \text{ and } \|f\|_{(c_0)'} = \|y\|_{\ell_1}.$$

The dual of ℓ^1 is isometrically isomorphic to ℓ^∞. The dual of $L^p([0, 1])$ is isometrically isomorphic to $L^q([0, 1])$ when $1 \leq p < \infty$ and $1/p + 1/q = 1$.

For every vector y in a Hilbert space H, the mapping

$$x \in H \to f_y(x) = \langle x, y \rangle$$

defines a continuous linear functional f_y on H. The Riesz representation theorem states that every continuous linear functional on H is of the form f_y for a uniquely defined vector y in H. The mapping $y \in H \to f_y$ is an antilinear isometric bijection from H onto its dual H'. When the scalars are real, this map is an isometric isomorphism.

When K is a compact Hausdorff topological space, the dual $M(K)$ of $C(K)$ is the space of Radon measures in the sense of Bourbaki. The subset $P(K)$ of $M(K)$ consisting of non-negative measures of mass 1 (probability measures) is a convex w*-closed subset of the unit ball of $M(K)$. The extreme points of $P(K)$ are the Dirac measures on K. The set of Dirac measures on K, equipped with the w*-topology, is homeomorphic to K.

Banach–Stone Theorem. If K and L are compact Hausdorff spaces and if $C(K)$ and $C(L)$ are isometrically isomorphic, then the topological spaces K and L are homeomorphic.

The result has been extended by Amir and Cambern to the case when the multiplicative Banach–Mazur distance between $C(K)$ and $C(L)$ is < 2. The theorem is no longer true when the distance is $= 2$.

In the commutative Banach algebra $C(K)$, the maximal ideals are precisely kernels of Dirac mesures on K,

$$I_x = \ker \delta_x = \{f \in C(K) : f(x) = 0\}, \quad x \in K.$$

More generally, by the Gelfand–Mazur theorem, the maximal ideals of a unital commutative Banach algebra can be identified with its characters—not merely as sets but as topological spaces: the former with the hull-kernel topology and the latter with the w*-topology. In this identification, the maximal ideal space can be viewed as a w*-compact subset of the unit ball in the dual A'.

Theorem. If K is a compact Hausdorff space, then the maximal ideal space Ξ of the Banach algebra $C(K)$ is homeomorphic to K.

Not every unital commutative Banach algebra is of the form $C(K)$ for some compact Hausdorff space K. However, this statement holds if one places $C(K)$ in the smaller category of commutative C*-algebras. Gelfand's representation theorem for commutative C*-algebras states that every commutative unital C^*-algebra A is isometrically isomorphic to a $C(K)$ space.The Hausdorff compact space K here is again the maximal ideal space, also called the spectrum of A in the C*-algebra context.

Bidual

If X is a normed space, the (continuous) dual X'' of the dual X' is called **bidual**, or **second dual** of X. For every normed space X, there is a natural map,

$$\begin{cases} F_X : X \to X'' \\ F_X(x)(f) = f(x) \quad \forall x \in X, \forall f \in X' \end{cases}$$

This defines $F_X(x)$ as a continuous linear functional on X', i.e., an element of X''. The map $F_X : x \to F_X(x)$ is a linear map from X to X''. As a consequence of the existence of a norming functional f for every x in X, this map F_X is isometric, thus injective.

For example, the dual of $X = c_0$ is identified with ℓ^1, and the dual of ℓ^1 is identified with ℓ^∞, the space of bounded scalar sequences. Under these identifications, F_X is the inclusion map from c_0 to ℓ^∞. It is indeed isometric, but not onto.

If F_X is surjective, then the normed space X is called reflexive (see below). Being the dual of a normed space, the bidual X'' is complete, therefore, every reflexive normed space is a Banach space.

Using the isometric embedding F_X, it is customary to consider a normed space X as a subset of its bidual. When X is a Banach space, it is viewed as a closed linear subspace of X''. If $X$is not reflexive, the unit ball of X is a proper subset of the unit ball of X''. The Goldstine theorem states that the unit ball of a normed space is weakly*-dense in the unit ball of the bidual. In other words, for every x'' in the bidual, there exists a net $\{x_j\}$ in X so that

$$\sup_j \|x_j\| \le \|x''\|, \quad x''(f) = \lim_j f(x_j), \quad f \in X'.$$

The net may be replaced by a weakly*-convergent sequence when the dual X' is separable. On the other hand, elements of the bidual of ℓ^1 that are not in ℓ^1 cannot be weak*-limit of *sequences* in ℓ^1, since ℓ^1 is weakly sequentially complete.

Banach's Theorems

Here are the main general results about Banach spaces that go back to the time of Banach's book (Banach (1932)) and are related to the Baire category theorem. According to this theorem, a complete metric space (such as a Banach space, a Fréchet space or an F-space) cannot be equal to a union of countably many closed subsets with empty interiors. Therefore, a Banach space cannot be the union of countably many closed subspaces, unless it is already equal to one of them; a Banach space with a countable Hamel basis is finite-dimensional.

Banach–Steinhaus Theorem. Let X be a Banach space and Y be a normed vector space. Suppose that F is a collection of continuous linear operators from X to Y. The uniform boundedness principle states that if for all x in X we have $\sup_{T\in F} \|T(x)\|_Y < \infty$, then $\sup_{T\in F} \|T\|_Y < \infty$.

The Banach–Steinhaus theorem is not limited to Banach spaces. It can be extended for example to the case where X is a Fréchet space, provided the conclusion is modified as follows: under the same hypothesis, there exists a neighborhood U of $\mathbf{0}$ in X such that all T in F are uniformly bounded on U,

$$\sup_{T\in F}\sup_{x\in U}\|T(x)\|_Y < \infty.$$

The Open Mapping Theorem. Let X and Y be Banach spaces and $T : X \to Y$ be a continuous linear operator. Then T is surjective if and only if T is an open map.

Corollary. Every one-to-one bounded linear operator from a Banach space onto a Banach space is an isomorphism.

The First Isomorphism Theorem for Banach spaces. Suppose that X and Y are Banach spaces and that $T \in B(X, Y)$. Suppose further that the range of T is closed in Y. Then $X/\mathrm{Ker}(T)$ is isomorphic to $T(X)$.

This result is a direct consequence of the preceding *Banach isomorphism theorem* and of the canonical factorization of bounded linear maps.

Corollary. If a Banach space X is the internal direct sum of closed subspaces M_1, ..., M_n, then X is isomorphic to $M_1 \oplus ... \oplus M_n$.

This is another consequence of Banach's isomorphism theorem, applied to the continuous bijection from $M_1 \oplus ... \oplus M_n$ onto X sending $(m_1, ..., m_n)$ to the sum $m_1 + ... + m_n$.

The Closed Graph Theorem. Let $T : X \to Y$ be a linear mapping between Banach spaces. The graph of T is closed in $X \times Y$ if and only if T is continuous.

Reflexivity

The normed space X is called **reflexive** when the natural map

$$\begin{cases} F_X : X \to X'' \\ F_X(x)(f) = f(x) \quad \forall x \in X, \forall f \in X' \end{cases}$$

is surjective. Reflexive normed spaces are Banach spaces.

Theorem. If X is a reflexive Banach space, every closed subspace of X and every quotient space of X are reflexive.

This is a consequence of the Hahn–Banach theorem. Further, by the open mapping theorem, if there is a bounded linear operator from the Banach space X onto the Banach space Y, then Y is reflexive.

Theorem. If X is a Banach space, then X is reflexive if and only if X' is reflexive.

Corollary. Let X be a reflexive Banach space. Then X is separable if and only if X' is separable.

Indeed, if the dual Y' of a Banach space Y is separable, then Y is separable. If X is reflexive and separable, then the dual of X' is separable, so X' is separable.

Theorem. Suppose that X_1, ..., X_n are normed spaces and that $X = X_1 \oplus ... \oplus X_n$. Then X is reflexive if and only if each X_j is reflexive.

Hilbert spaces are reflexive. The L^p spaces are reflexive when $1 < p < \infty$. More generally, uniformly convex spaces are reflexive, by the Milman–Pettis theorem. The spaces c_0, ℓ^1, $L^1([0, 1])$, $C([0, 1])$ are not reflexive. In these examples of non-reflexive spaces X, the bidual X'' is "much larger" than X. Namely, under the natural isometric embedding of X into X'' given by the Hahn–Banach theorem, the quotient X''/X is infinite-dimensional, and even nonseparable. However, Robert C. James has constructed an example of a non-reflexive space, usually called "*the James space*" and denoted by J, such that the quotient J''/J is one-dimensional. Furthermore, this space J is isometrically isomorphic to its bidual.

Theorem. A Banach space X is reflexive if and only if its unit ball is compact in the weak topology.

When X is reflexive, it follows that all closed and bounded convex subsets of X are weakly compact. In a Hilbert space H, the weak compactness of the unit ball is very often used in the following way: every bounded sequence in H has weakly convergent subsequences.

Weak compactness of the unit ball provides a tool for finding solutions in reflexive spaces to certain optimization problems. For example, every convex continuous function on the unit ball B of a reflexive space attains its minimum at some point in B.

As a special case of the preceding result, when X is a reflexive space over $\mathbf{R}$, every continuous linear functional f in X' attains its maximum $\|f\|$ on the unit ball of X. The following theorem of Robert C. James provides a converse statement.

James' Theorem. For a Banach space the following two properties are equivalent:

- X is reflexive.
- for all f in X' there exists x in X with $\|x\| \leq 1$, so that $f(x) = \|f\|$.

The theorem can be extended to give a characterization of weakly compact convex sets.

On every non-reflexive Banach space X, there exist continuous linear functionals that are not *norm-attaining*. However, the Bishop–Phelps theorem states that norm-attaining functionals are norm dense in the dual X' of X.

Weak Convergences of Sequences

A sequence $\{x_n\}$ in a Banach space X is weakly convergent to a vector $x \in X$ if $f(x_n)$ converges to $f(x)$ for every continuous linear functional f in the dual X'. The sequence $\{x_n\}$ is a weakly Cauchy sequence if $f(x_n)$ converges to a scalar limit $L(f)$, for every f in X'. A sequence $\{f_n\}$ in the dual X' is weakly* convergent to a functional $f \in X'$ if $f_n(x)$ converges to $f(x)$ for every x in X. Weakly Cauchy sequences, weakly convergent and weakly* convergent sequences are norm bounded, as a consequence of the Banach–Steinhaus theorem.

When the sequence $\{x_n\}$ in X is a weakly Cauchy sequence, the limit L above defines a bounded linear functional on the dual X', i.e., an element L of the bidual of X, and L is the limit of $\{x_n\}$ in the weak*-topology of the bidual. The Banach space X is weakly sequentially complete if every weakly Cauchy sequence is weakly convergent in X. It follows from the preceding discussion that reflexive spaces are weakly sequentially complete.

Theorem. For every measure μ, the space $L^1(\mu)$ is weakly sequentially complete.

An orthonormal sequence in a Hilbert space is a simple example of a weakly convergent sequence, with limit equal to the **0** vector. The unit vector basis of ℓ^p, $1 < p < \infty$, or of c_0, is another example of a **weakly null sequence**, i.e., a sequence that converges weakly to **0**. For every weakly null sequence in a Banach space, there exists a sequence of convex combinations of vectors from the given sequence that is norm-converging to **0**.

The unit vector basis of ℓ^1 is not weakly Cauchy. Weakly Cauchy sequences in ℓ^1 are weakly convergent, since L^1-spaces are weakly sequentially complete. Actually, weakly convergent sequences in ℓ^1 are norm convergent. This means that ℓ^1 satisfies Schur's property.

Results Involving the ℓ^1 Basis

Weakly Cauchy sequences and the ℓ^1 basis are the opposite cases of the dichotomy established in the following deep result of H. P. Rosenthal.

Theorem. Let $\{x_n\}$ be a bounded sequence in a Banach space. Either $\{x_n\}$ has a weakly Cauchy subsequence, or it admits a subsequence equivalent to the standard unit vector basis of ℓ^1.

A complement to this result is due to Odell and Rosenthal (1975).

Theorem. Let X be a separable Banach space. The following are equivalent:

- The space X does not contain a closed subspace isomorphic to ℓ^1.
- Every element of the bidual X'' is the weak*-limit of a sequence $\{x_n\}$ in X.

By the Goldstine theorem, every element of the unit ball B'' of X'' is weak*-limit of a net in the unit ball of X. When X does not contain ℓ^1, every element of B'' is weak*-limit of a *sequence* in the unit ball of X.

When the Banach space X is separable, the unit ball of the dual X', equipped with the weak*-topology, is a metrizable compact space K, and every element x'' in the bidual X''defines a bounded function on K:

$$x' \in K \mapsto x''(x'), \quad |x''(x')| \le \|x''\|.$$

This function is continuous for the compact topology of K if and only if x'' is actually in X, considered as subset of X''. Assume in addition for the rest of the paragraph that X does not contain ℓ^1. By the preceding result of Odell and Rosenthal, the function x'' is the pointwise limit on K of a sequence $\{x_n\} \subset X$ of continuous functions on K, it is therefore a first Baire class function on K. The unit ball of the bidual is a pointwise compact subset of the first Baire class on K.

Sequences, Weak And Weak* Compactness

When X is separable, the unit ball of the dual is weak*-compact by Banach–Alaoglu and metrizable for the weak* topology, hence every bounded sequence in the dual has weakly* convergent subsequences. This applies to separable reflexive spaces, but more is true in this case, as stated below.

The weak topology of a Banach space X is metrizable if and only if X is finite-dimensional. If the dual X' is separable, the weak topology of the unit ball of X is metrizable. This applies in particular to separable reflexive Banach spaces. Although the weak topology of the unit ball is not metrizable in general, one can characterize weak compactness using sequences.

Eberlein–**Šmulian** theorem. A set A in a Banach space is relatively weakly compact if and only if every sequence $\{a_n\}$ in A has a weakly convergent subsequence.

A Banach space X is reflexive if and only if each bounded sequence in X has a weakly convergent subsequence.

A weakly compact subset A in ℓ^1 is norm-compact. Indeed, every sequence in A has weakly convergent subsequences by Eberlein–Šmulian, that are norm convergent by the Schur property of ℓ^1.

Schauder Bases

A Schauder basis in a Banach space X is a sequence $\{e_n\}_{n\geq 0}$ of vectors in X with the property that for every vector x in X, there exist *uniquely* defined scalars $\{x_n\}_{n\geq 0}$ depending on x, such that

$$x = \sum_{n=0}^{\infty} x_n e_n, \quad i.e., \quad x = \lim_n P_n(x), \ P_n(x) := \sum_{k=0}^{n} x_k e_k.$$

Banach spaces with a Schauder basis are necessarily separable, because the countable set of finite linear combinations with rational coefficients (say) is dense.

It follows from the Banach–Steinhaus theorem that the linear mappings $\{P_n\}$ are uniformly bounded by some constant C. Let $\{e*n\}$ denote the coordinate functionals which assign to every x in X the coordinate x_n of x in the above expansion. They are called biorthogonal functionals. When the basis vectors have norm 1, the coordinate functionals $\{e*n\}$ have norm $\leq 2C$ in the dual of X.

Most classical separable spaces have explicit bases. The Haar system $\{h_n\}$ is a basis for $L^p([0, 1])$, $1 \leq p < \infty$. The trigonometric system is a basis in $L^p(\mathbf{T})$ when $1 < p < \infty$. The Schauder system is a basis in the space $C([0, 1])$. The question of whether the disk algebra $A(\mathbf{D})$ has a basis remained open for more than forty years, until Bočkarev showed in 1974 that $A(\mathbf{D})$ admits a basis constructed from the Franklin system.

Since every vector x in a Banach space X with a basis is the limit of $P_n(x)$, with P_n of finite rank and uniformly bounded, the space X satisfies the bounded approximation property. The first example by Enflo of a space failing the approximation property was at the same time the first example of a separable Banach space without a Schauder basis.

Robert C. James characterized reflexivity in Banach spaces with a basis: the space X with a Schauder basis is reflexive if and only if the basis is both shrinking and boundedly complete. In this case, the biorthogonal functionals form a basis of the dual of X.

Tensor Product

If $T_1 : X \times Y \to Z_1$ is any bilinear mapping into a **K**-vector space Z_1, then there exists a unique linear mapping $f : Z \to Z_1$ such that $T_1 = f \circ T$.

The image under T of a couple (x, y) in $X \times Y$ is denoted by $x \otimes y$, and called a simple tensor. Every element z in $X \otimes Y$ is a finite sum of such simple tensors.

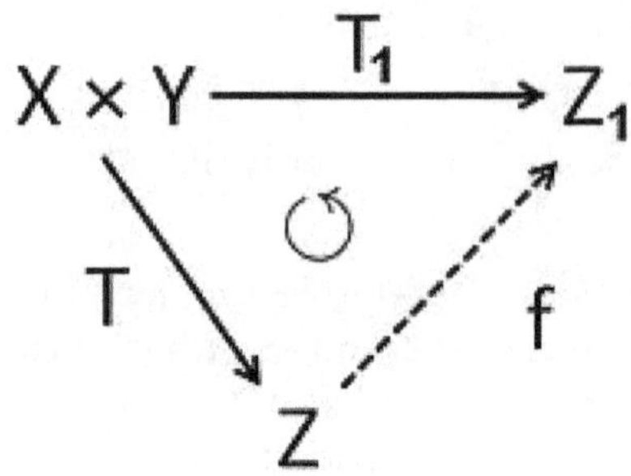

Let X and Y be two **K**-vector spaces. The tensor product $X \otimes Y$ of X and Y is a **K**-vector space Z with a bilinear mapping $T : X \times Y \to Z$ which has the following universal property:

There are various norms that can be placed on the tensor product of the underlying vector spaces, amongst others the projective cross norm and injective cross norm introduced by A. Grothendieck in 1955.

In general, the tensor product of complete spaces is not complete again. When working with Banach spaces, it is customary to say that the projective tensor product of two Banach spaces X and Y is the *completion* $X\widehat{\otimes}_\pi Y$ of the algebraic tensor product $X \otimes Y$ equipped with the projective tensor norm, and similarly for the injective tensor product $X\widehat{\otimes}_\varepsilon Y$. Grothendieck proved in particular that

$$C(K)\widehat{\otimes}_\varepsilon Y \simeq C(K,Y),$$
$$L^1([0,1])\widehat{\otimes}_\pi Y \simeq L^1([0,1],Y),$$

where K is a compact Hausdorff space, $C(K, Y)$ the Banach space of continuous functions from K to Y and $L^1([0, 1], Y)$ the space of Bochner-measurable and integrable functions from $[0, 1]$ to Y, and where the isomorphisms are isometric. The two isomorphisms above are the respective extensions of the map sending the tensor $f \otimes y$ to the vector-valued function $s \in K \to f(s)y \in Y$.

Tensor Products and the Approximation Property

Let X be a Banach space. The tensor product $X'\widehat{\otimes}_\varepsilon X$ is identified isometrically with the closure in $B(X)$ of the set of finite rank operators. When X has the approximation property, this closure coincides with the space of compact operators on X.

For every Banach space Y, there is a natural norm 1 linear map

$$Y\widehat{\otimes}_\pi X \to Y\widehat{\otimes}_\varepsilon X$$

obtained by extending the identity map of the algebraic tensor product. Grothendieck related the approximation problem to the question of whether this map is one-to-one when

Y is the dual of X. Precisely, for every Banach space X, the map

$$X'\widehat{\otimes}_\pi X \longrightarrow X'\widehat{\otimes}_\varepsilon X$$

is one-to-one if and only if X has the approximation property.

Grothendieck conjectured that $X\widehat{\otimes}_\pi Y$ and $X\widehat{\otimes}_\varepsilon Y$ must be different whenever X and Y are infinite-dimensional Banach spaces. This was disproved by Gilles Pisier in 1983. Pisier constructed an infinite-dimensional Banach space X such that $X\widehat{\otimes}_\pi X$ and $X\widehat{\otimes}_\varepsilon X$ are equal. Furthermore, just as Enflo's example, this space X is a "hand-made" space that fails to have the approximation property. On the other hand, Szankowski proved that the classical space $B(\ell^2)$ does not have the approximation property.

Some Classification Results

Characterizations of Hilbert Space among Banach Spaces

A necessary and sufficient condition for the norm of a Banach space X to be associated to an inner product is the parallelogram identity:

$$\forall x, y \in X: \qquad \|x+y\|^2 + \|x-y\|^2 = 2\left(\|x\|^2 + \|y\|^2\right).$$

It follows, for example, that the Lebesgue space $L^p([0, 1])$ is a Hilbert space only when $p = 2$. If this identity is satisfied, the associated inner product is given by the polarization identity. In the case of real scalars, this gives:

$$\langle x, y\rangle = \tfrac{1}{4}\left(\|x+y\|^2 - \|x-y\|^2\right).$$

For complex scalars, defining the inner product so as to be **C**-linear in x, antilinear in y, the polarization identity gives:

$$\langle x, y\rangle = \tfrac{1}{4}\left(\|x+y\|^2 - \|x-y\|^2 + i\left(\|x+iy\|^2 - \|x-iy\|^2\right)\right).$$

To see that the parallelogram law is sufficient, one observes in the real case that $<x, y>$ is symmetric, and in the complex case, that it satisfies the Hermitian symmetry property and $<ix, y> = i<x, y>$. The parallelogram law implies that $<x, y>$ is additive in x. It follows that it is linear over the rationals, thus linear by continuity.

Several characterizations of spaces isomorphic (rather than isometric) to Hilbert spaces are available. The parallelogram law can be extended to more than two vectors, and weakened by the introduction of a two-sided inequality with a constant $c \geq 1$: Kwapień proved that if

$$c^{-2}\sum_{k=1}^{n}\|x_k\|^2 \leq \operatorname{Ave}_{\pm}\left\|\sum_{k=1}^{n}\pm x_k\right\|^2 \leq c^2\sum_{k=1}^{n}\|x_k\|^2$$

for every integer n and all families of vectors $\{x_1, ..., x_n\} \subset X$, then the Banach space X is isomorphic to a Hilbert space. Here, $\text{Ave}_{\pm}$ denotes the average over the 2^n possible choices of signs ±1. In the same article, Kwapień proved that the validity of a Banach-valued Parseval's theorem for the Fourier transform characterizes Banach spaces isomorphic to Hilbert spaces.

Lindenstrauss and Tzafriri proved that a Banach space in which every closed linear subspace is complemented (that is, is the range of a bounded linear projection) is isomorphic to a Hilbert space. The proof rests upon Dvoretzky's theorem about Euclidean sections of high-dimensional centrally symmetric convex bodies. In other words, Dvoretzky's theorem states that for every integer n, any finite-dimensional normed space, with dimension sufficiently large compared to n, contains subspaces nearly isometric to the n-dimensional Euclidean space.

The next result gives the solution of the so-called *homogeneous space problem*. An infinite-dimensional Banach space X is said to be homogeneous if it is isomorphic to all its infinite-dimensional closed subspaces. A Banach space isomorphic to ℓ^2 is homogeneous, and Banach asked for the converse.

Theorem. A Banach space isomorphic to all its infinite-dimensional closed subspaces is isomorphic to a separable Hilbert space.

An infinite-dimensional Banach space is hereditarily indecomposable when no subspace of it can be isomorphic to the direct sum of two infinite-dimensional Banach spaces. The Gowers dichotomy theorem asserts that every infinite-dimensional Banach space X contains, either a subspace Y with unconditional basis, or a hereditarily indecomposable subspace Z, and in particular, Z is not isomorphic to its closed hyperplanes. If X is homogeneous, it must therefore have an unconditional basis. It follows then from the partial solution obtained by Komorowski and Tomczak–Jaegermann, for spaces with an unconditional basis, that X is isomorphic to ℓ^2.

Spaces of Continuous Functions

When two compact Hausdorff spaces K_1 and K_2 are homeomorphic, the Banach spaces $C(K_1)$ and $C(K_2)$ are isometric. Conversely, when K_1 is not homeomorphic to K_2, the (multiplicative) Banach–Mazur distance between $C(K_1)$ and $C(K_2)$ must be greater than or equal to 2, see above the results by Amir and Cambern. Although uncountable compact metric spaces can have different homeomorphy types, one has the following result due to Milutin:

Theorem. Let K be an uncountable compact metric space. Then $C(K)$ is isomorphic to $C([0, 1])$.

The situation is different for countably infinite compact Hausdorff spaces. Every countably infinite compact K is homeomorphic to some closed interval of ordinal numbers

$$\langle 1, \alpha \rangle = \{\gamma : 1 \leq \gamma \leq \alpha\}$$

equipped with the order topology, where is a countably infinite ordinal. The Banach space $C(K)$ is then isometric to C(<1, >). When , are two countably infinite ordinals, and assuming ≤ , the spaces *C*(<1, >) and *C*(<1, >) are isomorphic if and only if < . For example, the Banach spaces

$$C(\langle 1,\omega\rangle),\ C(\langle 1,\omega^{\omega}\rangle),\ C(\langle 1,\omega^{\omega^2}\rangle),\ C(\langle 1,\omega^{\omega^3}\rangle),\cdots,C(\langle 1,\omega^{\omega^{\omega}}\rangle),\cdots$$

are mutually non-isomorphic.

Examples

A glossary of symbols:

- **K** = **R**, **C**;
- *X* is a compact Hausdorff space;
- *I* is a closed and bounded interval [*a*, *b*];
- *p*, *q* are real numbers with $1 < p, q < \infty$ so that $1/p + 1/q = 1$.
- Σ is a σ-algebra of sets;
- Ξ is an algebra of sets (for spaces only requiring finite additivity, such as the ba space);
- *μ* is a measure with variation |*μ*|.

Classical Banach spaces					
	Dual space	**Reflexive**	**weakly sequentially complete**	**Norm**	**Notes**
$\mathbf{K}^n$	$\mathbf{K}^n$	Yes	Yes	$\|x\|_2 = \left(\sum_{i=1}^{n} \|x_i\|^2\right)^{\frac{1}{2}}$	Euclidean space
ℓ_p^n	ℓ_q^n	Yes	Yes	$\|x\|_p = \left(\sum_{i=1}^{n} \|x_i\|^p\right)^{\frac{1}{p}}$	
ℓ_∞^n	ℓ_1^n	Yes	Yes	$\|x\|_\infty = \max_{1\le i\le n} \|x_i\|$	
ℓ_p	ℓ_q	Yes	Yes	$\|x\|_p = \left(\sum_{i=1}^{\infty} \|x_i\|^p\right)^{\frac{1}{p}}$	
ℓ_1	ℓ_∞	No	Yes	$\|x\|_1 = \sum_{i=1}^{\infty} \|x_i\|$	
ℓ_∞	ba	No	No	$\|x\|_\infty = \sup_i \|x_i\|$	
c	ℓ_1	No	No	$\|x\|_\infty = \sup_i \|x_i\|$	
c_0	ℓ_1	No	No	$\|x\|_\infty = \sup_i \|x_i\|$	Isomorphic but not isometric to c.

c_0	ℓ_1	No	No	$\|x\|_\infty = \sup_i \|x_i\|$	Isomorphic but not isometric to c.
bv	ℓ_∞	No	Yes	$\|x\|_{bv} = \|x_1\| + \sum_{i=1}^{\infty} \|x_{i+1} - x_i\|$	Isometrically isomorphic to ℓ_1.
bv$_0$	ℓ_∞	No	Yes	$\|x\|_{bv_0} = \sum_{i=1}^{\infty} \|x_{i+1} - x_i\|$	Isometrically isomorphic to ℓ_1.
bs	ba	No	No	$\|x\|_{bs} = \sup_n \left\| \sum_{i=1}^{n} x_i \right\|$	Isometrically isomorphic to ℓ_∞.
cs	ℓ_1	No	No	$\|x\|_{bs} = \sup_n \left\| \sum_{i=1}^{n} x_i \right\|$	Isometrically isomorphic to c.
$B(X, \Xi)$	ba(Ξ)	No	No	$\|f\|_B = \sup_{x \in X} \|f(x)\|$	
$C(X)$	rca(X)	No	No	$\|x\|_{C(X)} = \max_{x \in X} \|f(x)\|$	
ba(Ξ)	?	No	Yes	$\|\mu\|_{ba} = \sup_{A \in \Sigma} \|\mu\|(A)$	
ca(Σ)	?	No	Yes	$\|\mu\|_{ba} = \sup_{A \in \Sigma} \|\mu\|(A)$	A closed subspace of ba(Σ).
rca(Σ)	?	No	Yes	$\|\mu\|_{ba} = \sup_{A \in \Sigma} \|\mu\|(A)$	A closed subspace of ca(Σ).
$L^p(\mu)$	$L^q(\mu)$	Yes	Yes	$\|f\|_p = \left(\int \|f\|^p \, d\mu \right)^{\frac{1}{p}}$	
$L^1(\mu)$	$L^\infty(\mu)$	No	Yes	$\|f\|_1 = \int \|f\| \, d\mu$	The dual is $L^\infty(\mu)$ if μ is σ-finite.
BV(I)	?	No	Yes	$\|f\|_{BV} = V_f(I) + \lim_{x \to a^+} f(x)$	$V_f(I)$ is the total variation of f
NBV(I)	?	No	Yes	$\|f\|_{BV} = V_f(I)$	NBV(I) consists of BV(I) functions such that $\lim_{x \to a^+} f(x) = 0$
AC(I)	$\mathbf{K} + L^\infty(I)$	No	Yes	$\|f\|_{BV} = V_f(I) + \lim_{x \to a^+} f(x)$	Isomorphic to the Sobolev space $W^{1,1}(I)$.
$C^n([a, b])$	rca($[a,b]$)	No	No	$\|f\| = \sum_{i=0}^{n} \sup_{x \in [a,b]} \|f^{(i)}(x)\|$	Isomorphic to $\mathbf{R}^n \oplus C([a,b])$, essentially by Taylor's theorem.

Derivatives

Several concepts of a derivative may be defined on a Banach space. The Fréchet derivative allows for an extension of the concept of a directional derivative to Banach spaces. The Gâteaux derivative allows for an extension of a directional derivative to locally convex topological vector spaces. Fréchet differentiability is a stronger condition than Gâteaux differentiability. The quasi-derivative is another generalization of directional derivative that implies a stronger condition than Gâteaux differentiability, but a weaker condition than Fréchet differentiability.

Generalizations

Several important spaces in functional analysis, for instance the space of all infinitely often differentiable functions **R** → **R**, or the space of all distributions on **R**, are complete but are not normed vector spaces and hence not Banach spaces. In Fréchet spaces one still has a complete metric, while LF-spaces are complete uniform vector spaces arising as limits of Fréchet spaces.

F-Space

In functional analysis, an F-space is a vector space V over the real or complex numbers together with a metric $d : V \times V \to \mathrm{R}$ so that

- Scalar multiplication in V is continuous with respect to d and the standard metric on **R** or **C**.
- Addition in V is continuous with respect to d.
- The metric is translation-invariant; i.e., $d(x + a, y + a) = d(x, y)$ for all x, y and a in V
- The metric space (V, d) is complete

Some authors call these spaces *Fréchet spaces*, but usually the term is reserved for locally convex F-spaces. The metric may or may not necessarily be part of the structure on an F-space; many authors only require that such a space be metrizable in a manner that satisfies the above properties.

REFERENCES

1. MSC classification, 2010, available at http://www.ams.org/msc/msc2010.html, Classification 11HXX.
2. Schmidt's books. Grötschel et alii, Lovász et alii, Lovász.
3. Cassels (1971) p. 203
4. Grötschel et alii, Lovász et alii, Lovász, and Beck and Robins.
5. Schmidt, Wolfgang M. Norm form equations. Ann. Math. (2) 96 (1972), pp. 526-551. See also Schmidt's books; compare Bombieri and Vaaler and also Bombieri and Gubler.
6. For Kolmogorov's normability theorem, see Walter Rudin's Functional Analysis. For more results, see Schneider, and Thompson and see Kalton et alii.
7. Kalton et alii. Gardner
8. Enrico Bombieri and Walter Gubler (2006). Heights in Diophantine Geometry. Cambridge U. P.

9. J. W. S. Cassels. An Introduction to the Geometry of Numbers. Springer Classics in Mathematics, Springer-Verlag 1997 (reprint of 1959 and 1971 Springer-Verlag editions).
10. John Horton Conway and N. J. A. Sloane, Sphere Packings, Lattices and Groups, Springer-Verlag, NY, 3rd ed., 1998.
11. Hancock, Harris (1939). Development of the Minkowski Geometry of Numbers. Macmillan. (Republished in 1964 by Dover.)
12. Hazewinkel, Michiel, ed. (2001), "Geometry of numbers", Encyclopedia of Mathematics, Springer, ISBN 978-1-55608-010-4
13. Edmund Hlawka, Johannes Schoißengeier, Rudolf Taschner. Geometric and Analytic Number Theory. Universitext. Springer-Verlag, 1991.
14. C. G. Lekkerkerker. Geometry of Numbers. Wolters-Noordhoff, North Holland, Wiley. 1969.
15. Wolfgang M. Schmidt. Diophantine approximation. Lecture Notes in Mathematics 785. Springer. (1980 [1996 with minor corrections])
16. Wolfgang M. Schmidt.Diophantine approximations and Diophantine equations, Lecture Notes in Mathematics, Springer Verlag 2000.
17. Siegel, Carl Ludwig (1989). Lectures on the Geometry of Numbers. Springer-Verlag.
18. Rolf Schneider, Convex bodies: the Brunn-Minkowski theory, Cambridge University Press, Cambridge, 1993.
19. "Minkowski's theorem". PlanetMath.
20. Stevenhagen, Peter. Number Rings.
21. Malyshev, A.V. (2001), "Minkowski theorem", in Hazewinkel, Michiel, Encyclopedia of Mathematics, Springer, ISBN 978-1-55608-010-4
22. Bombieri, Enrico; Gubler, Walter (2006). Heights in Diophantine Geometry. New Mathematical Monographs. 4. Cambridge: Cambridge University Press. doi:10.2277/0521846153. ISBN 978-0-521-71229-3. MR 2216774. Zbl 1130.11034.
23. Schlickewei, Hans Peter (1977). "On norm form equations". J. Number Theory. 9 (3): 370–380. doi:10.1016/0022-314X(77)90072-5. MR 0444562.
24. Schmidt, Wolfgang M. (1972). "Norm form equations". Annals of Mathematics. Second Series. 96 (3): 526–551. doi:10.2307/1970824. MR 0314761.
25. Schmidt, Wolfgang M. (1980). Diophantine approximation. Lecture Notes in Mathematics. 785 (1996 with minor corrections ed.). Berlin: Springer-Verlag. doi:10.1007/978-3-540-38645-2. ISBN 3-540-09762-7. MR 568710. Zbl 0421.10019.
26. Schmidt, Wolfgang M. (1991). Diophantine approximations and Diophantine equations. Lecture Notes in Mathematics. 1467. Berlin: Springer-Verlag. doi:10.1007/BFb0098246. ISBN 3-540-54058-X. MR 1176315. Zbl 0754.11020.
27. Callier, Frank M. (1991). Linear System Theory. New York: Springer-Verlag.

ISBN 0-387-97573-X.

28. Kedlaya, Kiran S. (2010), p-adic differential equations, Cambridge Studies in Advanced Mathematics, 125, Cambridge University Press, ISBN 978-0-521-76879-5, Theorem 1.3.6
29. Rolewicz, Stefan (1987), Functional analysis and control theory: Linear systems, Mathematics and its Applications (East European Series), 29 (Translated from the Polish by Ewa Bednarczuk ed.), Dordrecht; Warsaw: D. Reidel Publishing Co.; PWN—Polish Scientific Publishers, pp. xvi+524, ISBN 90-277-2186-6, MR 920371, OCLC 13064804
30. Banach, Stefan (1932), Théorie des opérations linéaires (PDF), Monografie Matematyczne, 1, Warszawa: Subwencji Funduszu Kultury Narodowej, Zbl 0005.20901.
31. Beauzamy, Bernard (1985) [1982], Introduction to Banach Spaces and their Geometry (Second revised ed.), North-Holland.
32. Bourbaki, Nicolas (1987), Topological vector spaces, Elements of mathematics, Berlin: Springer-Verlag, ISBN 978-3-540-13627-9.
33. Carothers, Neal L. (2005), A short course on Banach space theory, London Mathematical Society Student Texts, 64, Cambridge: Cambridge University Press, pp. xii+184, ISBN 0-521-84283-2.
34. Diestel, Joseph (1984), Sequences and series in Banach spaces, Graduate Texts in Mathematics, 92, New York: Springer-Verlag, pp. xii+261, ISBN 0-387-90859-5.
35. Dunford, Nelson; Schwartz, Jacob T. with the assistance of W. G. Bade and R. G. Bartle (1958), Linear Operators. I. General Theory, Pure and Applied Mathematics, 7, New York: Interscience Publishers, Inc., MR 0117523
36. Lindenstrauss, Joram; Tzafriri, Lior (1977), Classical Banach Spaces I, Sequence Spaces, Ergebnisse der Mathematik und ihrer Grenzgebiete, 92, Berlin: Springer-Verlag, ISBN 3-540-08072-4.
37. Megginson, Robert E. (1998), An introduction to Banach space theory, Graduate Texts in Mathematics, 183, New York: Springer-Verlag, pp. xx+596, ISBN 0-387-98431-3.
38. Ryan, Raymond A. (2002), Introduction to Tensor Products of Banach Spaces, Springer Monographs in Mathematics, London: Springer-Verlag, pp. xiv+225, ISBN 1-85233-437-1.
39. Wojtaszczyk, Przemysław (1991), Banach spaces for analysts, Cambridge Studies in Advanced Mathematics, 25, Cambridge: Cambridge University Press, pp. xiv+382, ISBN 0-521-35618-0.

CHAPTER 6

Computational Number Theory

Chapter Contents

Computational number theory is the branch of number theory concerned with finding and implementing efficient computer algorithms for solving various problems in number theory. Much progress has been made in this field in recent years, both in terms of improved computer speed and in terms of finding more efficient algorithms. Two important applications of computational number theory are primality testing and prime factorization of large integers.

Primality testing is considered easy in the sense that very large general numbers (currently up to 4000 digits or so) can be tested reliably for primality. In fact, on August 6, 2002, Agrawal, Saxena, and Kayal found a polynomial time algorithm for testing and proving the primality of general numbers. Although this algorithm is still impractical, it was a landmark discovery, since polynomial time algorithms are considered easy. On the other hand, factoring is considered hard in the sense that no polynomial time algorithm is currently known for factoring integers. The largest general integer to be factored was RSA-576, a 174-digit number that is the product of two 87-digit primes. The fact that primality testing is easy but factoring is hard allows for secure encryption, such as RSA encryption.

Other problems in computational number theory include computing the greatest common divisor of large numbers and computing various quantities associated with number fields, i.e., class numbers and class groups.

INTRODUCTION

In mathematics and computer science, computational number theory, also known as algorithmic number theory, is the study of algorithms for performing number theoreticcomputations.

COMPUTATIONAL COMPLEXITY OF MATHEMATICAL OPERATIONS

The following tables list the running time of various algorithms for common mathematical operations.

Here, complexity refers to the time complexity of performing computations on a multitape Turing machine. See big O notation for an explanation of the notation used.

Note: Due to the variety of multiplication algorithms, $M(n)$ below stands in for the complexity of the chosen multiplication algorithm.

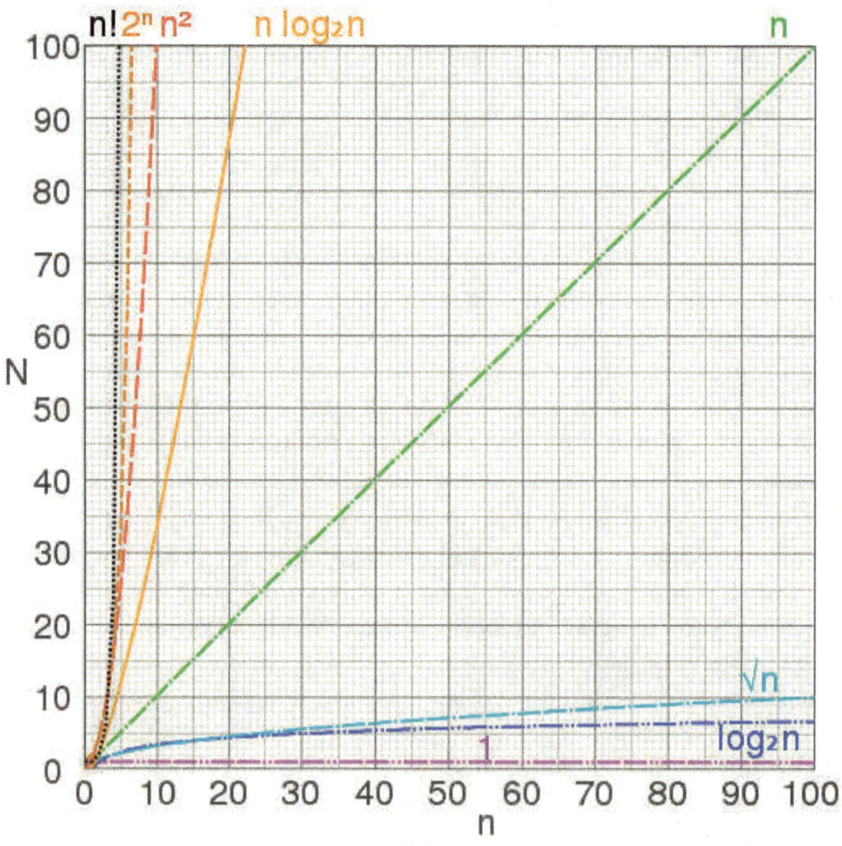

Graphs of number of operations, N vs input size, n for common complexities, assuming a coefficient of 1.

Arithmetic Functions

Operation	Input	Output	Algorithm	Complexity
Addition	Two n-digit numbers N, N	One n+1-digit number	Schoolbook addition with carry	$\Theta(n)$, $\Theta(log(N))$
Subtraction	Two n-digit numbers N, N	One n+1-digit number	Schoolbook subtraction with borrow	$\Theta(n)$, $\Theta(log(N))$
Multiplication	Two n-digit numbers	One $2n$-digit number	Schoolbook long multiplication	$O(n^2)$
			Karatsuba algorithm	$O(n^{1.585})$
			3-way Toom–Cook multiplication	$O(n^{1.465})$
			k-way Toom–Cook multiplication	$O(n^{\log(2k-1)/\log k})$
			Mixed-level Toom–Cook (Knuth 4.3.3-T)	$O(n\ 2^{\sqrt{2 \log n}} \log n)$
			Schönhage–Strassen algorithm	$O(n \log n \log \log n)$
			Fürer's algorithm	$O(n \log n\ 2^{O(\log^* n)})$
Division	Two n-digit numbers	One n-digit number	Schoolbook long division	$O(n^2)$
			Newton–Raphson division	$O(M(n))$
Square root	One n-digit number	One n-digit number	Newton's method	$O(M(n))$
Modular exponentiation	Two n-digit numbers and a k-bit exponent	One n-digit number	Repeated multiplication and reduction	$O(M(n)\ 2^k)$
			Exponentiation by squaring	$O(M(n)\ k)$
			Exponentiation with Montgomery reduction	$O(M(n)\ k)$

Algebraic Functions

Operation	Input	Output	Algorithm	Complexity
Polynomial evaluation	One polynomial of degree n with fixed-size polynomial coefficients	One fixed-size number	Direct evaluation	$\Theta(n)$
			Horner's method	$\Theta(n)$

Polynomial gcd (over $\mathbf{Z}[x]$ or $F[x]$)	Two polynomials of degree n with fixed-size polynomial coefficients	One polynomial of degree at most n	Euclidean algorithm	$O(n^2)$
			Fast Euclidean algorithm	$O(M(n) \log n)$

Special Functions

Many of the methods in this section are given in Borwein & Borwein.

Elementary Functions

The elementary functions are constructed by composing arithmetic operations, the exponential function (exp), the natural logarithm (log), trigonometric functions (sin, cos), and their inverses. The complexity of an elementary function is equivalent to that of its inverse, since all elementary functions are analytic and hence invertible by means of Newton's method. In particular, if either exp or log in the complex domain can be computed with some complexity, then that complexity is attainable for all other elementary functions.

Below, the size n refers to the number of digits of precision at which the function is to be evaluated.

Algorithm	Applicability	Complexity
Taylor series; repeated argument reduction (e.g. $\exp(2x) = [\exp(x)]^2$) and direct summation	exp, log, sin, cos, arctan	$O(M(n)\, n^{1/2})$
Taylor series; FFT-based acceleration	exp, log, sin, cos, arctan	$O(M(n)\, n^{1/3} (\log n)^2)$
Taylor series; binary splitting + bit burst method	exp, log, sin, cos, arctan	$O(M(n) (\log n)^2)$
Arithmetic-geometric mean iteration	exp, log, sin, cos, arctan	$O(M(n) \log n)$

It is not known whether $O(M(n) \log n)$ is the optimal complexity for elementary functions. The best known lower bound is the trivial bound $\Omega(M(n))$.

Non-Elementary Functions

Function	Input	Algorithm	Complexity
Gamma function	n-digit number	Series approximation of the incomplete gamma function	$O(M(n)\, n^{1/2} (\log n)^2)$
	Fixed rational number	Hypergeometric series	$O(M(n) (\log n)^2)$
	$m/24$, m an integer	Arithmetic-geometric mean iteration	$O(M(n) \log n)$

Hypergeometric function ${}_pF_q$	n-digit number	(As described in Borwein & Borwein)	$O(M(n)\, n^{1/2} (\log n)^2)$
	Fixed rational number	Hypergeometric series	$O(M(n) (\log n)^2)$

Mathematical Constants

This table gives the complexity of computing approximations to the given constants to n correct digits.

Constant	Algorithm	Complexity
Golden ratio, φ	Newton's method	$O(M(n))$
Square root of 2, $\sqrt{2}$	Newton's method	$O(M(n))$
Euler's number, e	Binary splitting of the Taylor series for the exponential function	$O(M(n) \log n)$
	Newton inversion of the natural logarithm	$O(M(n) \log n)$
Pi, π	Binary splitting of the arctan series in Machin's formula	$O(M(n) (\log n)^2)$
	Salamin–Brent algorithm	$O(M(n) \log n)$
Euler's constant, γ	Sweeney's method (approximation in terms of the exponential integral)	$O(M(n) (\log n)^2)$

Number Theory

Algorithms for number theoretical calculations are studied in computational number theory.

Operation	Input	Output	Algorithm	Complexity
Greatest common divisor	Two n-digit numbers	One number with at most n digits	Euclidean algorithm	$O(n^2)$
			Binary GCD algorithm	$O(n^2)$
			Left/right k-ary binary GCD algorithm	$O(n^2/ \log n)$
			Stehlé–Zimmermann algorithm	$O(M(n) \log n)$
			Schönhage controlled Euclidean descent algorithm	$O(M(n) \log n)$
Jacobi symbol	Two n-digit numbers	0, −1, or 1		
			Schönhage controlled Euclidean descent algorithm	$O(M(n) \log n)$
			Stehlé–Zimmermann algorithm	$O(M(n) \log n)$

Factorial	A fixed-size number m	One $O(m \log m)$-digit number	Bottom-up multiplication	$O(m^2 \log m)$
			Binary splitting	$O(M(m \log m) \log m)$
			Exponentiation of the prime factors of m	$O(M(m \log m) \log \log m)$, $O(M(m \log m))$

MATRIX ALGEBRA

The following complexity figures assume that arithmetic with individual elements has complexity $O(1)$, as is the case with fixed-precision floating-point arithmetic or operations on a finite field.

Operation	Input	Output	Algorithm	Complexity
Matrix multiplication	Two $n \times n$ matrices	One $n \times n$ matrix	Schoolbook matrix multiplication	$O(n^3)$
			Strassen algorithm	$O(n^{2.807})$
			Coppersmith–Winograd algorithm	$O(n^{2.376})$
			Optimized CW-like algorithms	$O(n^{2.373})$
Matrix multiplication	One $n \times m$ matrix & one $m \times p$ matrix	One $n \times p$ matrix	Schoolbook matrix multiplication	$O(nmp)$
Matrix inversion*	One $n \times n$ matrix	One $n \times n$ matrix	Gauss–Jordan elimination	$O(n^3)$
			Strassen algorithm	$O(n^{2.807})$
			Coppersmith–Winograd algorithm	$O(n^{2.376})$
			Optimized CW-like algorithms	$O(n^{2.373})$
Singular value decomposition	One $m \times n$ matrix	One mxm matrix, one mxn matrix, & one nxn matrix		$O(mn^2)$ $(m \leq n)$
		One mxr matrix, one rxr matrix, & one nxr matrix		

Determinant	One $n \times n$ matrix	One number	Laplace expansion	$O(n!)$
			Division-free algorithm	$O(n^4)$
			LU decomposition	$O(n^3)$
			Bareiss algorithm	$O(n^3)$
			Fast matrix multiplication	$O(n^{2.373})$
Back substitution	Triangular matrix	n solutions	Back substitution	$O(n^2)$

In 2005, Henry Cohn, Robert Kleinberg, Balázs Szegedy, and Chris Umans showed that either of two different conjectures would imply that the exponent of matrix multiplication is 2.

* Because of the possibility of blockwise inverting a matrix, where an inversion of an $n \times n$ matrix requires inversion of two half-sized matrices and six multiplications between two half-sized matrices, and since matrix multiplication has a lower bound of $\Omega(n^2 \log n)$ operations, it can be shown that a divide and conquer algorithm that uses blockwise inversion to invert a matrix runs with the same time complexity as the matrix multiplication algorithm that is used internally.

SageMath

SageMath (previously Sage or SAGE, System for Algebra and Geometry Experimentation) is mathematical software with features covering many aspects of mathematics, including algebra, combinatorics, numerical mathematics, number theory, and calculus.

The first version of SageMath was released on 24 February 2005 as free and open source software under the terms of the GNU General Public License version 2, with the initial goals of creating an «open source alternative to Magma, Maple, Mathematica, and MATLAB». The originator and leader of the SageMath project, William Stein, is a mathematician at the University of Washington.

SageMath «uses a Python-like syntax,» supporting procedural, functional and object-oriented constructs.

Features

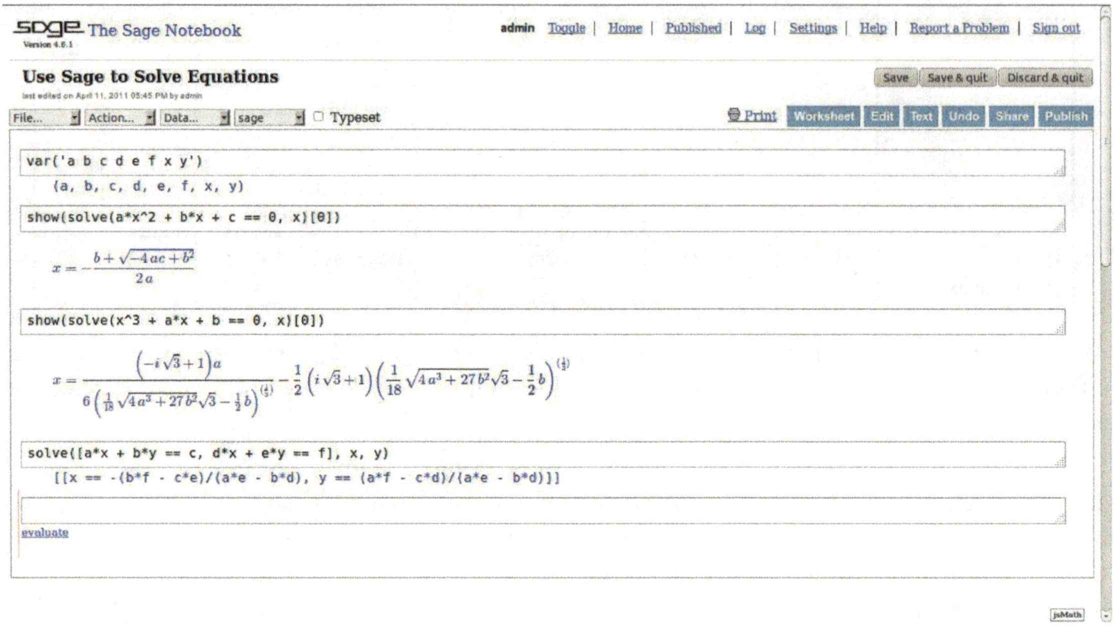

Equation solving and typesetting using the SageMath notebook web interface

Features of SageMath include:

- A browser-based notebook for review and re-use of previous inputs and outputs, including graphics and text annotations. Compatible with Firefox, Opera, Konqueror, Google Chrome and Safari. Notebooks can be accessed locally or remotely and the connection can be secured with HTTPS.
- A text-based command-line interface using IPython
- Support for parallel processing using multi-core processors, multiple processors, or distributed computing
- Calculus using Maxima and SymPy
- Numerical linear algebra using the GSL, SciPy and NumPy
- Libraries of elementary and special mathematical functions
- 2D and 3D graphs of symbolic functions and numerical data
- Matrix manipulation, including sparse arrays
- Multivariate statistics libraries, using R and SciPy
- A toolkit for adding user interfaces to calculations and applications
- Graph theory visualization and analysis tools
- Libraries of number theory functions
- Support for complex numbers, arbitrary precision and symbolic computation

- Technical word processing including formula editing and embedding SageMath within LaTeX documents
- The Python standard library, including tools for connecting to SQL, HTTP, HTTPS, NNTP, IMAP, SSH, IRC, FTP and others
- Interfaces to some third-party applications like Mathematica, Magma, R, and Maple
- MoinMoin as a Wiki system for knowledge management
- Documentation using Sphinx
- An automated test-suite
- Execution of Fortran, C, C++, and Cython code
- Although not provided by SageMath directly, SageMath can be called from within Mathematica; as is done in this Mathematica notebook example

Development

William Stein realized when designing Sage that there were many open-source mathematics software packages already written in different languages, namely C, C++, Common Lisp, Fortran and Python.

Rather than reinventing the wheel, Sage (which is written mostly in Python and Cython) integrates many specialized mathematics software packages into a common interface, for which a user needs to know only Python. However, Sage contains hundreds of thousands of unique lines of code adding new functions and creating the interface between its components.

William A. Stein

SageMath uses both students and professionals for development. The development of SageMath is supported by both volunteer work and grants. However, it was not until 2016 that the first full-time Sage developer was hired (funded by an EU grant). The same year,

Stein described his disappointment with a lack of academic funding and credentials for software development, citing it as the reason for his decision to leave his tenured academic position to work full-time on the project in a newly founded company, SageMath, Inc.

Release History

Only the major releases are listed below. SageMath practices the "release early, release often" concept, with releases every few weeks or months. In total, there have been over 300 releases, although their frequency has decreased.

SageMath versions			
Version	**Release Date**	**Description**	
0.1	January 2005	Included PARI, but not GAP or Singular	
0.2–0.4	March to July 2005	Cremona's database, multivariate polynomials, large finite fields and more documentation	
0.5–0.7	August to September 2005	Vector spaces, rings, modular symbols, and windows usage	
0.8	October 2005	Full distribution of GAP, Singular	
0.9	November 2005	Maxima and clisp added	
1.0	February 2006		
2.0	January 2007		
3.0	April 2008	Interacts, R interface	
4.0	May 2009	Solaris 10 support, 64-bit OS X support	
5.0	May 2012	OS X Lion support	
6.0	December 2013	SageMath Development moved to Git	
7.0	January 2016	Massive GUI improvement	

Achievements

- 2007: first prize in the scientific software division of Les Trophées du Libre, an international competition for free software.
- 2012: one of the projects selected for the Google Summer of Code.
- 2013: ACM/SIGSAM Jenks Prize.
- SageMath has been cited in a variety of publications.

Performance

Both binaries and source code are available for SageMath from the download page. If SageMath is built from source code, many of the included libraries such as ATLAS, FLINT, and NTL will be tuned and optimized for that computer, taking into account the number of processors, the size of their caches, whether there is hardware support for SSE instructions, etc.

Cython can increase the speed of SageMath programs, as the Python code is converted into C.

Licensing and Availability

SageMath is free software, distributed under the terms of the GNU General Public License version 2+. SageMath is available in many ways:

- The source code can be downloaded from the downloads page. Although not recommended for end users, development releases of SageMath are also available.
- Binaries can be downloaded for Linux, macOS and Solaris (both x86 and SPARC).
- A live CD containing a bootable Linux operating system is also available. This allows usage of SageMath without Linux installation.
- Users could use an online version of SageMath at sagenb.org, but it has been discontinued in April 2015.
- Users can use an online "single cell" version of SageMath at sagecell.sagemath.org or embed a single SageMath cell into any web page. Users can also create permalinks to SageMath computations using the cell server.
- A new online SageMath notebook is available at cloud.sagemath.com.

Although Microsoft was sponsoring a native version of SageMath for the Windows operating system, as of 2012 there were no plans for a native port, and users of Windows currently have to use virtualization technology such as VirtualBox to run SageMath. As of SageMath 5.9, it mostly successfully builds on Cygwin.

Linux distributions in which SageMath is available as a package are Mandriva, Fedora, and Arch Linux. It is also available as a dedicated Ubuntu PPA. In Gentoo, it's available via layman in the «sage-on-gentoo» overlay. However, SageMath can be installed to any Linux distribution.

Gentoo prefix also provides Sage on other operating systems.

Software Packages Contained in SageMath

The philosophy of SageMath is to use existing open-source libraries wherever they exist. Therefore, it uses many libraries from other projects.

Mathematics packages contained in SageMath	Algebra	GAP, Maxima, Singular
	Algebraic geometry	Singular
	Arbitrary precision arithmetic	MPIR, MPFR, MPFI, NTL, mpmath
	Arithmetic geometry	PARI/GP, NTL, mwrank, ECM
	Calculus	Maxima, SymPy, GiNaC
	Combinatorics	Symmetrica, Sage-Combinat
	Linear algebra	ATLAS, BLAS, LAPACK, NumPy, LinBox, IML, GSL
	Graph theory	NetworkX
	Group theory	GAP
	Numerical computation	GSL, SciPy, NumPy, ATLAS
	Number theory	PARI/GP, FLINT, NTL
	Statistical computing	R, SciPy
Other packages contained in SageMath	Command-line shell	IPython
	Database	ZODB, SQLite
	Graphical interface	SageMath Notebook, jsMath
	Graphics	matplotlib, Tachyon, GD, Jmol
	Interactive programming language	Python
	Networking	Twisted
Other Mathematics package available for SageMath	Differential Geometry and Tensor Calculus	Sage Manifolds

Usage Examples

Algebra and Calculus

```
x, a, b, c = var('x, a, b, c')
# Note that IPython also supports a faster way to do this, by calling
# this equivalent expression starting with a comma:
# ,var x a b c

log(sqrt(a)).simplify_log() # returns 1/2*log(a)
log(a / b).expand_log() # returns log(a) - log(b)
sin(a + b).simplify_trig() # returns sin(a)*cos(b) + sin(b)*cos(a)
cos(a + b).simplify_trig() # returns -sin(a)*sin(b) + cos(a)*cos(b)
(a + b)^5 # returns (a + b)^5
expand((a + b) ^ 5) # a^5 + 5*a^4*b + 10*a^3*b^2 + 10*a^2*b^3 + 5*a*b^4 + b^5

limit((x ^ 2 + 1) / (2 + x + 3 * x ^ 2), x=Infinity) # returns 1/3
limit(sin(x) / x, x=0) # returns 1

diff(acos(x), x) # returns -1/sqrt(-x^2 + 1)
f = exp(x) * log(x)
f.diff(x, 3) # returns e^x*log(x) + 3*e^x/x - 3*e^x/x^2 + 2*e^x/x^3

solve(a * x ^ 2 + b * x + c, x) # returns [x == -1/2*(b + sqrt(-4*a*c + b^2))/a,
                                #          x == -1/2*(b - sqrt(-4*a*c + b^2))/a]
```

```
f = x ^ 2 + 432 / x
solve(f.diff(x) == 0, x) # returns [x == 3*I*sqrt(3) - 3,
                          # x == -3*I*sqrt(3) - 3, x == 6]
```

Differential Equations

```
t = var('t') # define a variable t
x = function('x', t) # define x to be a function of that variable
de = (diff(x, t) + x == 1)
desolve(de, [x, t]) # returns (c + e^t)*e^(-t)
```

Linear Algebra

```
A = matrix([[1, 2, 3], [3, 2, 1], [1, 1, 1]])
y = vector([0, -4, -1])
A.solve_right(y) # returns (-2, 1, 0)
A.eigenvalues() # returns [5, 0, -1]

B = matrix([[1, 2, 3], [3, 2, 1], [1, 2, 1]])
B.inverse() # returns
  [   0  1/2 -1/2]
  [-1/4 -1/4    1]
  [ 1/2    0 -1/2]

# same matrix, but over the ring of doubles (not rationals, as above)
sage: B = matrix(RDF, [[1, 2, 3], [3, 2, 1], [1, 2, 1]])
sage: B.inverse()

[-5.55111512313e-17          0.5         -0.5]
[             -0.25        -0.25          1.0]
[               0.5          0.0         -0.5]

# Call NumPy for the Moore-Penrose pseudo-inverse,
# since SageMath does not support that yet.

import numpy
C = matrix([[1 , 1], [2 , 2]])
matrix(numpy.linalg.pinv(C)) # returns
  [0.1 0.2]
  [0.1 0.2]
```

Number Theory

```
prime_pi(1000000) # returns 78498, the number of primes less than one million

E = EllipticCurve('389a') # construct an elliptic curve from its Cremona label
P, Q = E.gens()
7 * P + Q # returns (2418773145843925 3/244328192262001 :
        # 3778434777075334029261244/3819094217575529893001 : 1)
sage: E2 = EllipticCurve(CC, [0,0,-2,1,1])
sage: E2
Elliptic Curve defined by y^2 + (-2.00000000000000)*y =
        x^3 + 1.00000000000000*x + 1.00000000000000 over
        Complex Field with 53 bits of precision
sage: E2.j_invariant()
61.7142857142857
```

REFERENCES

1. A. Schönhage, A.F.W. Grotefeld, E. Vetter: Fast Algorithms—A Multitape Turing Machine Implementation, BI Wissenschafts-Verlag, Mannheim, 1994
2. D. Knuth. The Art of Computer Programming, Volume 2. Third Edition, Addison-Wesley 1997.
3. Martin Fürer. Faster Integer Multiplication. Proceedings of the 39th Annual ACM Symposium on Theory of Computing, San Diego, California, USA, June 11–13, 2007, pp. 55–67.
4. http://planetmath.org/fasteuclideanalgorithm
5. J. Borwein & P. Borwein. Pi and the AGM: A Study in Analytic Number Theory and Computational Complexity. John Wiley 1987.
6. David and Gregory Chudnovsky. Approximations and complex multiplication according to Ramanujan. Ramanujan revisited, Academic Press, 1988, pp 375–472.
7. Richard P. Brent, Multiple-precision zero-finding methods and the complexity of elementary function evaluation, in: Analytic Computational Complexity (J. F. Traub, ed.), Academic Press, New York, 1975, 151–176.
8. J. Sorenson. (1994). "Two Fast GCD Algorithms". Journal of Algorithms. 16 (1): 110–144. doi:10.1006/jagm.1994.1006.
9. R. Crandall & C. Pomerance. Prime Numbers – A Computational Perspective. Second Edition, Springer 2005.
10. Möller N (2008). "On Schönhage's algorithm and subquadratic integer gcd computation" (PDF). Mathematics of Computation. 77 (261): 589–607. doi:10.1090/S0025-5718-07-02017-0.
11. Bernstein D J. "Faster Algorithms to Find Non-squares Modulo Worst-case Integers".

12. Richard P. Brent; Paul Zimmermann (2010). "An O(M(n) log n) algorithm for the Jacobi symbol". arXiv:1004.2091Freely accessible.
13. P. Borwein. "On the complexity of calculating factorials". Journal of Algorithms 6, 376-380 (1985)
14. Davie, A.M.; Stothers, A.J. (2013), "Improved bound for complexity of matrix multiplication", Proceedings of the Royal Society of Edinburgh, 143A: 351–370, doi:10.1017/S0308210511001648
15. Vassilevska Williams, Virginia (2011), Breaking the Coppersmith-Winograd barrier (PDF)
16. Le Gall, François (2014), "Powers of tensors and fast matrix multiplication", Proceedings of the 39th International Symposium on Symbolic and Algebraic Computation (ISSAC 2014), arXiv:1401.7714Freely accessible
17. http://page.mi.fu-berlin.de/rote/Papers/pdf/Division-free+algorithms.pdf
18. Aho, Alfred V.; Hopcroft, John E.; Ullman, Jeffrey D. (1974), The Design and Analysis of Computer Algorithms, Addison-Wesley, Theorem 6.6, p. 241
19. J. B. Fraleigh and R. A. Beauregard, "Linear Algebra," Addison-Wesley Publishing Company, 1987, p 95.
20. Henry Cohn, Robert Kleinberg, Balazs Szegedy, and Chris Umans. Group-theoretic Algorithms for Matrix Multiplication. arXiv:math.GR/0511460. Proceedings of the 46th Annual Symposium on Foundations of Computer Science, 23–25 October 2005, Pittsburgh, PA, IEEE Computer Society, pp. 379–388.
21. Ran Raz. On the complexity of matrix product. In Proceedings of the thirty-fourth annual ACM symposium on Theory of computing. ACM Press, 2002. doi:10.1145/509907.509932.
22. T.H. Cormen, C.E. Leiserson, R.L. Rivest, C. Stein, Introduction to Algorithms, 3rd ed., MIT Press, Cambridge, MA, 2009, §28.2.
23. https://git.sagemath.org/sage.git/tree/README.md
24. Stein, William. "SAGE: A Computer System for Algebra and Geometry Experimentation". Retrieved 30 March 2012.
25. Stein, William (2007-06-12). "Sage Days 4" (PDF). Archived from the original (PDF) on 2007-06-27. Retrieved 2007-08-02.
26. Anastassiou, George A.; Mezei, Razvan A. (2015). Numerical Analysis Using Sage. New York: Springer. pp. xl and 1. ISBN 9783319167381.
27. "Sage documentation".
28. "SageMath Interact functionality". Retrieved 2008-04-11.
29. "Using SageTeX".
30. "Using Compiled Code Interactively". SageMath Documentation. Retrieved 14 July 2011.
31. "Calling SageMath from Mathematica".
32. "days7 - Sage Wiki". Wiki.sagemath.org. 2008-11-14. Retrieved 2013-12-09.

33. "Sage – Acknowledgement". Retrieved 2010-07-13.
34. William Stein: The origins of SageMath – creating a viable open source alternative to Magma, Maple, Mathematica, and Matlab (presentation, June 11, 2016)
35. "SageMath Download - src-old". Retrieved 17 July 2011.
36. "sage-5.0.txt". Retrieved 17 May 2012.
37. "Installing and using SageMath just got even easier". Retrieved 12 July 2014.
38. "Free Software Brings Affordability, Transparency To Mathematics". Science Daily. December 7, 2007. Retrieved 2008-07-20.
39. "Sage Mathematical Software System". Retrieved 9 June 2012.
40. "SIGSAM: Awards and prizes". Retrieved 2 Aug 2013.
41. "Publications Citing Sage". Retrieved 14 July 2011.
42. "Publications Citing Sage-Combinat". Retrieved 14 July 2011.
43. "Cython, Sage, and the Need for Speed".
44. "About the SageMath Cell Server". Sagecell.sagemath.org. Retrieved 2013-12-09.
45. "Sage – Acknowledgment".
46. Stein, William (16 March 2012). "Re: Question about Sage". Google Groups. sage-devel. Retrieved March 18, 2012.
47. "Information for the port to Cygwin". Retrieved 12 June 2013.
48. "Sage : "AIMS" team".
49. "sage-on-gentoo".
50. "SageMath Standard Packages". Retrieved 10 June 2012.
51. "SageManifolds: home".

CHAPTER 7

Transcendental Number Theory

Chapter Contents

A transcendental number is a (possibly complex) number that is not the root of *any* integer polynomial, meaning that it is not an algebraic number of any degree. Every real transcendental number must also be irrational, since a rational number is, by definition, analgebraic number of degree one.

A complex number z can be tested to see if it is transcendental using the Wolfram Language command Not[Element[*x*, Algebraics]].

Transcendental numbers are important in the history of mathematics because their investigation provided the first proof that circle squaring, one of the geometric problems of antiquity that had baffled mathematicians for more than 2000 years was, in fact, insoluble. Specifically, in order for a number to be produced by a geometric construction using the ancient Greek rules, it must be either rational or a very special kind of algebraic number known as a Euclidean number. Because the number π is transcendental, the construction cannot be done according to the Greek rules.

Liouville showed how to construct special cases (such as Liouville's constant) using Liouville's approximation theorem. In particular, he showed that any number that has a rapidly converging sequence of rational approximations must be transcendental. For many years, it was only known how to determine if special classes of numbers were transcendental. The determination of the status of more general numbers was considered an important enough unsolved problem that it was one of Hilbert's problems.

INTRODUCTION

Transcendental number theory is a branch of number theory that investigates transcendental numbers, in both qualitative and quantitative ways.

The set of transcendental numbers is uncountably infinite. Since the polynomials with rational coefficients are countable, and since each such polynomial has a finite number of zeroes, the algebraic numbers must also be countable. However, Cantor's diagonal argument proves that the real numbers (and therefore also the complex numbers) are uncountable. Since the real numbers are the union of algebraic and transcendental numbers, they cannot both be countable. This makes the transcendental numbers uncountable.

TRANSCENDENCE

The fundamental theorem of algebra tells us that if we have a non-zero polynomial with integer coefficients then that polynomial will have a root in the complex numbers. That is, for any polynomial P with integer coefficients there will be a complex number α such that $P(\alpha) = 0$. Transcendence theory is concerned with the converse question, given a complex number α, is there a polynomial P with integer coefficients such that $P(\alpha) = 0$? If no such polynomial exists then the number is called transcendental.

More generally the theory deals with algebraic independence of numbers. A set of numbers $\{\alpha_1,\alpha_2,\ldots,\alpha_n\}$ is called algebraically independent over a field k if there is no non-zero polynomial P in n variables with coefficients in k such that $P(\alpha_1,\alpha_2,\ldots,\alpha_n) = 0$. So working out if a given number is transcendental is really a special case of algebraic independence where our set consists of just one number.

A related but broader notion than "algebraic" is whether there is a closed-form expression for a number, including exponentials and logarithms as well as algebraic operations. There are various definitions of "closed-form", and questions about closed-form can often be reduced to questions about transcendence.

HISTORY

Approximation by Rational Numbers: Liouville to Roth

Use of the term *transcendental* to refer to an object that is not algebraic dates back to the seventeenth century, when Gottfried Leibniz proved that the sine function was not an algebraic function. The question of whether certain classes of numbers could be transcendental dates back to 1748 when Euler asserted that the number $\log_a b$ was not algebraic for rational numbers a and b provided b is not of the form $b = a^c$ for some rational c.

Euler's assertion was not proved until the twentieth century, but almost a hundred years after his claim Joseph Liouville did manage to prove the existence of numbers that are not algebraic, something that until then had not been known for sure. His original papers on the matter in the 1840s sketched out arguments using continued fractions to construct transcendental numbers. Later, in the 1850s, he gave a necessary condition for a number to be algebraic, and thus a sufficient condition for a number to be transcendental. This transcendence criterion was not strong enough to be necessary too, and indeed it fails to detect that the number e is transcendental. But his work did provide a larger class of transcendental numbers, now known as Liouville numbers in his honour.

Liouville's criterion essentially said that algebraic numbers cannot be very well approximated by rational numbers. So if a number can be very well approximated by rational numbers then it must be transcendental. The exact meaning of "very well approximated" in Liouville's work relates to a certain exponent. He showed that if α is an algebraic number of degree $d \geq 2$ and ε is any number greater than zero, then the expression

$$\left|\alpha - \frac{p}{q}\right| < \frac{1}{q^{d+\varepsilon}}$$

can be satisfied by only finitely many rational numbers p/q. Using this as a criterion for transcendence is not trivial, as one must check there are infinitely many solutions p/q for every $d \geq 2$.

In the twentieth century work by Axel Thue, Carl Siegel, and Klaus Roth reduced the exponent in Liouville's work from $d + \varepsilon$ to $d/2 + 1 + \varepsilon$, and finally, in 1955, to $2 + \varepsilon$. This result, known as the Thue–Siegel–Roth theorem, is ostensibly best possible, since if the exponent $2 + \varepsilon$ is replaced by just 2 then the result is no longer true. However, Serge Lang conjectured an improvement of Roth's result; in particular he conjectured that $q^{2+\varepsilon}$ in the denominator of the right-hand side could be reduced to $q^2\log(q)^{1+\varepsilon}$.

Roth's work effectively ended the work started by Liouville, and his theorem allowed mathematicians to prove the transcendence of many more numbers, such as the Champernowne constant. The theorem is still not strong enough to detect *all* transcendental numbers, though, and many famous constants including e and π either are not or are not known to be very well approximable in the above sense.

Auxiliary Functions: Hermite to Baker

Fortunately other methods were pioneered in the nineteenth century to deal with the algebraic properties of e, and consequently of π through Euler's identity. This work centred on use of the so-called auxiliary function. These are functions which typically have many zeros at the points under consideration. Here "many zeros" may mean literally a lot of zeros, or as few as one zero but with a high multiplicity, or even many zeros all with high multiplicity. Charles Hermite used auxiliary functions that approximated the functions e^{kx} for each natural number k in order to prove the transcendence of e in 1873. His work was built upon by Ferdinand von Lindemann in the 1880s in order to prove that e^{α} is transcendental for nonzero algebraic numbers α. In particular this proved that π is transcendental since $e^{\pi i}$ is algebraic, and thus answered the problem of antiquity as to whether it was possible to square the circle. Karl Weierstrass developed their work yet further and eventually proved the Lindemann–Weierstrass theorem in 1885.

In 1900 David Hilbert posed his famous collection of problems. The seventh of these, and one of the hardest in Hilbert's estimation, asked about the transcendence of numbers of the form a^b where a and b are algebraic, a isn't zero or one, and b is irrational. In the 1930s Alexander Gelfond and Theodor Schneider proved that all such numbers were indeed transcendental using a non-explicit auxiliary function whose existence was granted by Siegel's lemma. This result, the Gelfond–Schneider theorem, proved the transcendence of numbers such as e^π and the Gelfond–Schneider constant.

The next big result in this field occurred in the 1960s, when Alan Baker made progress on a problem posed by Gelfond on linear forms in logarithms. Gelfond himself had managed to find a non-trivial lower bound for the quantity

$$|\beta_1 \log \alpha_1 + \beta_2 \log \alpha_2|$$

where all four unknowns are algebraic, the αs being neither zero nor one and the βs being irrational. Finding similar lower bounds for the sum of three or more logarithms had eluded Gelfond, though. The proof of Baker's theorem contained such bounds, solving Gauss' class number problem for class number one in the process. This work won Baker the Fields medal for its uses in solving Diophantine equations. From a purely transcendental number theoretic viewpoint, Baker had proved that if $\alpha_1,\ldots,\alpha_n$ are algebraic numbers, none of them zero or one, and $\beta_1,\ldots,\beta_n$ are algebraic numbers such that $1,\beta_1,\ldots,\beta_n$ are linearly independent over the rational numbers, then the number

$$\alpha_1^{\beta_1}\alpha_2^{\beta_2}\cdots\alpha_n^{\beta_n}$$

is transcendental.

Other Techniques: Cantor and Zilber

In the 1870s Georg Cantor started to develop set theory and in 1874 published a paper proving that the algebraic numbers could be put in one-to-one correspondence with the set of natural numbers, and thus that the set of transcendental numbers must be uncountable. Later, in 1891, Cantor used his more familiar diagonal argument to prove the same result. While Cantor's result is often quoted as being purely existential and thus unusable for constructing a single transcendental number, the proofs in both the aforementioned papers give methods to construct transcendental numbers.

While Cantor used set theory to prove the plenitude of transcendental numbers, a recent development has been the use of model theory in attempts to prove an unsolved problem in transcendental number theory. The problem is to determine the transcendence degree of the field

$$K = \mathbb{Q}(x_1,\ldots,x_n,e^{x_1},\ldots,e^{x_n})$$

for complex numbers $x_1,\ldots,x_n$ that are linearly independent over the rational numbers. Stephen Schanuel conjectured that the answer is at least n, but no proof is known. In 2004, though, Boris Zilber published a paper that used model theoretic techniques to create a structure that behaves very much like the complex numbers equipped with the operations of addition, multiplication, and exponentiation. Moreover, in this abstract structure Schanuel's conjecture does indeed hold. Unfortunately it is not yet known that this structure is in fact the same as the complex numbers with the operations mentioned, it could be that Schanuel's conjecture is false and that there exists some other abstract structure that behaves very similarly to the complex numbers but where Schanuel's conjecture holds. Zilber did provide several criteria that would prove the structure in question was **C**, but could not prove the so-called Strong Exponential Closure axiom. The simplest case of this axiom has since been proved, but a proof that it holds in full generality is required to complete the proof of the conjecture.

APPROACHES

A typical problem in this area of mathematics is to work out whether a given number is transcendental. Cantor used a cardinality argument to show that there are only countably many algebraic numbers, and hence almost all numbers are transcendental. Transcendental numbers therefore represent the typical case; even so, it may be extremely difficult to prove that a given number is transcendental (or even simply irrational).

For this reason transcendence theory often works towards a more quantitative approach. So given a particular complex number α one can ask how close α is to being an algebraic number. For example, if one supposes that the number α is algebraic then can one show that it must have very high degree or a minimum polynomial with very large coefficients? Ultimately if it is possible to show that no finite degree or size of coefficient is sufficient then the number must be transcendental. Since a number α is transcendental if and only if $P(\alpha) \neq 0$ for every non-zero polynomial P with integer coefficients, this problem can be approached by trying to find lower bounds of the form

$$|P(a)| > F(A, d)$$

where the right hand side is some positive function depending on some measure A of the size of the coefficients of P, and its degree d, and such that these lower bounds apply to all $P \neq 0$. Such a bound is called a transcendence measure.

The case of $d = 1$ is that of "classical" diophantine approximation asking for lower bounds for

$$|ax + b|.$$

The methods of transcendence theory and diophantine approximation have much in common: they both use the auxiliary function concept.

MAJOR RESULTS

The Gelfond–Schneider theorem was the major advance in transcendence theory in the period 1900–1950. In the 1960s the method of Alan Baker on linear forms in logarithms of algebraic numbers reanimated transcendence theory, with applications to numerous classical problems and diophantine equations.

OPEN PROBLEMS

While the Gelfond–Schneider theorem proved that a large class of numbers was transcendental, this class was still countable. Many well known mathematical constants are still not known to be transcendental, and in some cases it is not even known whether they are rational or irrational. A partial list can be found here.

A major problem in transcendence theory is showing that a particular set of numbers is algebraically independent rather than just showing that individual elements are transcendental. So while we know that e and ϖ are transcendental that doesn't imply that $e + \varpi$ is transcendental, nor other combinations of the two (except e^{π}, Gelfond's constant, which is known to be transcendental). Another major problem is dealing with numbers that are not related to the exponential function. The main results in transcendence theory tend to revolve around e and the logarithm function, which means that wholly new methods tend to be required to deal with numbers that cannot be expressed in terms of these two objects in an elementary fashion.

Schanuel's conjecture would solve the first of these problems somewhat as it deals with algebraic independence and would indeed confirm that $e+\varpi$ is transcendental. It still revolves around the exponential function however and so would not necessarily deal with numbers such as Apéry's constant or the Euler–Mascheroni constant. Another extremely difficult unsolved problem is the so-called Constant or Identity problem.

REFERENCES

1. N. Bourbaki, Elements of the History of Mathematics Springer (1994).
2. Euler, L. (1748). Introductio in analysin infinitorum. Lausanne.
3. J. Liouville, Sur les classes très étendues de quantités dont la valeur n'est ni algébrique ni même réductible à des irrationelles algébriques, Comptes Rendus Acad. Sci. Paris 18, (1844), pp.883–885, 910–911; Journal Math. Pures et Appl. 16, (1851), pp.133–142.
4. Thue, A. (1909). "Über Annäherungswerte algebraischer Zahlen". J. Reine Angew. Math. 135: 284–305. doi:10.1515/crll.1909.135.284.
5. Siegel, C. L. (1921). "Approximation algebraischer Zahlen". Math. Zeitschrift. 10 (3–4): 172–213. doi:10.1007/BF01211608.
6. Roth, K. F. (1955). "Rational approximations to algebraic numbers". Mathematika. 2 (1): 1–20. doi:10.1112/S0025579300000644. And "Corrigendum", p. 168, doi:10.1112/S0025579300000826.
7. Mahler, K. (1953). "On the approximation of π". Proc. Akad. Wetensch. Ser. A. 56: 30–42.
8. Hermite, C. (1873). "Sur la fonction exponentielle". C. R. Acad. Sci. Paris. 77.
9. Lindemann, F. (1882). "Ueber die Zahl π ". Mathematische Annalen. 20 (2): 213–225. doi:10.1007/BF01446522.
10. Weierstrass, K. (1885). "Zu Hrn. Lindemann's Abhandlung: ‹Über die Ludolph'sche Zahl›". Sitzungber. Königl. Preuss. Akad. Wissensch. zu Berlin.

2: 1067–1086.

11. Gelfond, A. O. (1934). "Sur le septième Problème de D. Hilbert". Izv. Akad. Nauk SSSR. 7: 623–630.
12. Schneider, T. (1935). "Transzendenzuntersuchungen periodischer Funktionen. I. Transzendend von Potenzen". J. Reine Angew. Math. 172: 65–69. doi:10.1515/crll.1935.172.65.
13. A.Baker, Linear forms in the logarithms of algebraic numbers. I, II, III, Mathematika 13 ,(1966), pp.204–216; ibid. 14, (1967), pp.102–107; ibid. 14, (1967), pp.220–228, MR0220680
14. Cantor, G. (1874). "Ueber eine Eigenschaft des Ingebriffes aller reelen algebraischen Zahlen". J. Reine Angew. Math. (in German). 77: 258–262. doi:10.1515/crll.1874.77.258.
15. Cantor, G. (1891). "Ueber eine elementare Frage der Mannigfaltigkeitslehre". Jahresbericht der Deutschen Mathematiker-Vereinigung (in German). 1: 75–78.
16. Kac, M.; Stanislaw, U. (1968). Mathematics and Logic. Fredering A. Praeger. p. 13.
17. Bell, E. T. (1937). Men of Mathematics. New York: Simon & Schuster. p. 569.
18. Gray, R. (1994). "Georg Cantor and Transcendental Numbers" (PDF). Amer. Math. Monthly. 101 (9): 819–832. JSTOR 2975129.
19. Zilber, B. (2005). "Pseudo-exponentiation on algebraically closed fields of characteristic zero". Annals of Pure and Applied Logic. 132 (1): 67–95. doi:10.1016/j.apal.2004.07.001. MR 2102856.
20. Marker, D. (2006). "A remark on Zilber's pseudoexponentiation". J. Symbolic Logic. 71 (3): 791–798. doi:10.2178/jsl/1154698577. JSTOR 27588482. MR 2250821.
21. Richardson, D. (1968). "Some Undecidable Problems Involving Elementary Functions of a Real Variable". J. Symbolic Logic. 33 (4): 514–520. JSTOR 2271358. MR 0239976.

Index